CW01185467

THE *YOGASŪTRA* OF PATAÑJALI

This book offers a systematic and radical introduction to the Buddhist roots of Pātañjala-yoga, or the Yoga system of Patañjali. By examining each of the 195 aphorisms (*sūtras*) of the *Yogasūtra* and discussing the *Yogabhāṣya*, it shows that traditional and popular views on Pātañjala-yoga obscure its true nature. The book argues that Patañjali's Yoga contains elements rooted in both orthodox and heterodox philosophical traditions, including Sāṅkhya, Jaina and Buddhist thought.

With a fresh translation and a detailed commentary on the *Yogasūtra*, the author unearths how several of the terms, concepts and doctrines in Patañjali's Yoga can be traced to Buddhism, particularly the Abhidharma Buddhism of Vasubandhu and the early Yogācāra of Asaṅga. The work presents the *Yogasūtra* of Patañjali as a synthesis of two perspectives: the metaphysical perspective of Sāṅkhya and the empirical–psychological perspective of Buddhism. Based on a holistic understanding of Yoga, the study explores key themes of the text, such as meditative absorption, means, supernormal powers, isolation, Buddhist conceptions of meditation and the interplay between Sāṅkhya and Buddhist approaches to suffering and emancipation. It further highlights several new findings and clarifications on textual interpretation and discrepancies.

An important intervention in Indian and Buddhist philosophy, this book opens up a new way of looking at the Yoga of Patañjali in the light of Buddhism beyond standard approaches and will greatly interest scholars and researchers of Buddhist studies, Yoga studies, Indian philosophy, philosophy in general, literature, religion and comparative studies, Indian and South Asian studies and the history of ideas.

Pradeep P. Gokhale is Honorary Adjunct Professor in the Departments of Philosophy and Pali, Savitribai Phule Pune University, Pune, India. He has thirty-one years of postgraduate teaching and research experience at Savitribai Phule Pune University and was Dr B. R. Ambedkar Research Professor at the Central Institute of Higher Tibetan Studies, Sarnath (Varanasi), for six years. He has written in diverse areas such as Indian epistemology and logic; schools of classical Indian philosophy such as Buddhism, Lokāyata, Yoga and Jainism; Indian moral philosophy, social philosophy, Indian philosophy of religion and contemporary Buddhism. His research interests have focused on the interface between orthodox and heterodox Indian thought, including between Nyāya and Buddhist logical thought.

THE *YOGASŪTRA* OF PATAÑJALI

A New Introduction to the Buddhist Roots of the Yoga System

Pradeep P. Gokhale

Routledge
Taylor & Francis Group
LONDON AND NEW YORK

First published 2020
by Routledge
2 Park Square, Milton Park, Abingdon, Oxon OX14 4RN

and by Routledge
52 Vanderbilt Avenue, New York, NY 10017

Routledge is an imprint of the Taylor & Francis Group, an informa business

British Library Cataloguing-in-Publication Data
A catalogue record for this book is available from the British Library

Library of Congress Cataloging-in-Publication Data
A catalog record has been requested for this book

ISBN: 978-0-367-40898-5 (hbk)
ISBN: 978-0-367-45670-2 (pbk)
ISBN: 978-0-367-81595-0 (ebk)

Typeset in Bembo
by Apex CoVantage, LLC

CONTENTS

List of illustrations *vi*
Acknowledgements *viii*

Introduction 1

1 On meditative absorption (*Samādhipādaḥ*) 19

2 On means (*Sādhanapādaḥ*) 68

3 On supernormal powers (*Vibhūtipādaḥ*) 116

4 On isolation (*Kaivalyapādaḥ*) 161

5 Concluding observations 188

Appendix I: Buddhist conceptions of meditation, yoga and bodhisattva's *spiritual journey vis-à-vis Pātañjala-yoga* *199*
*Appendix II: Asaṅga on forms of supernormal knowledge and powers (*abhijñās *and* ṛddhis*)* *212*
Appendix III: Discrepancies between Patañjali's aphorisms and Vyāsa's interpretations *219*
Glossary *222*
Aids to reading romanized Sanskrit *232*
Bibliography *233*
Index *237*

ILLUSTRATIONS

Figure

2.1 Vasubandhu's classification of karma 76

Tables

1.1 Vasubandhu's classification of mind (*citta*) into two groups: defiled (*kliṣṭa*) and wholesome (*kuśala*) 21
1.2 Phenomenological translation of Sāṅkhya terminology 24
1.3 The stages in *rūpadhyāna* 34
1.4 The stages of *samprajñāta-samādhi* 35
1.5 Vasubandhu and Vyāsa on *vitarka* and *vicāra* 36
1.6 Blissful births caused by the stages of meditation 37
2.1 Types of defilements according to Pātañjala-yoga and Buddhism 71
2.2 Four foundational categories according to Buddhism and Yoga 80
2.3 Seven types of knowledge according to Abhidharma Buddhism 87
2.4 Four types of knowledge designating emancipation according to Vyāsa and Vasubandhu 88
2.5 Types of causes according to Vyāsa and Asaṅga 90
2.6 Additional eleven types in Asaṅga's list 91
2.7 *Vrata* (Jainism), *Yama* (Yoga) and *Śīla* (Buddhism) 92
2.8 The nature of *yamas*/*vratas* according to Vyāsa and Umāsvāti 93
3.1 Three temporal paths according to Vyāsa and Vasubandhu 126
3.2 The correlation between the objects of meditation and supernormal powers according to Patañjali 135
3.3 Lower worlds according to Vyāsa and Vasubandhu 136

3.4 Middle world (between Naraka and *svarga*) 136
3.5 Description of the *svarga* on Sumeru 136
3.6 The lowest *svarga* as the *svarga* of enjoyment and the kinds of gods who reside in it 137
3.7 Gods residing in higher *svargas* according to Vasubandhu and Vyāsa 138
3.8 Abodes of gods belonging to the third (highest?) *svarga* and the objects of their meditational enjoyment according to Vyāsa's account 138
3.9 Eight supernormal powers 149
3.10 Bodily wealth and beauty as a supernormal achievement 150
3.11 The common pattern followed in two aphorisms 151
3.12 Classification of yoga-practitioners according to the two traditions 154
3.13 Ordered sequence of direct realizations (*abhisamaya-krama*) 155
4.1 *Siddhis* (Yoga) and *ṛddhis* (Buddhism) 162
4.2 Creation of minds through meditation 164
4.3 Classification of causes according to Buddhism and Patañjali 169
4.4 Types of questions according to Buddhism and Vyāsa 183
A1.1 Four *dhyānas* and their aspects 201
A1.2 Four *dhyānas* according to Asaṅga and Vasubandhu 201
A1.3 Correlation between thirteen *vihāras* and ten *bhūmis* 208
A3.1 Discrepancies between *Yogasūtra* and *Yogabhāṣya* 220

ACKNOWLEDGEMENTS

I first presented the seminal idea behind this book in "The Thought Construction in *Yogasūtra*", a paper included in the anthology *Yoga: Its Philosophy and Science* (published by P. D. Ghanekar of Datta Lakshmi Trust, Pune, in 1995). I was able to develop the idea in part because my friend, Professor Shrikant Bahulkar, supplied me with de La Vallée-Poussin's "Le Bouddhisme et le Yoga de Patañjali", and another friend, Dr S. M. Bhave, translated the paper for me from French into Marathi. Later, I pursued the idea in collaboration with Professors Mahesh Deokar and Shrikant Bahulkar as well as other colleagues in the Department of Pali at Savitribai Phule Pune University. In June 2012, a workshop was held in the Department of Pali on "Pātañjala-Yoga and Buddhism", in which we collectively read some pages of *Yogasūtra* and *Yogabhāṣya* in the light of Buddhism. I benefitted immensely from the proceedings of the workshop, and although we were unable to continue the project collectively, I pursued it as an individual project. Professors Bahulkar and Deokar encouraged me and were always helpful in my endeavour. A revised version of my 1995 paper was published in 2015 as "Interplay of Sāṅkhya and Buddhist Ideas in the Yoga of Patañjali (With Special Reference to *Yogasūtra* and *Yogabhāṣya*)" in the *Journal of Buddhist Studies*, Sri Lanka, in the publication of which Ven. Dhammajoti's editorial suggestions proved very helpful.

This led me to search more rigorously for connections between the *Yogasūtra* and the Buddhist Abhidharma and Yogācāra texts. My search was greatly aided by the *Index to the Abhidharmakośabhāṣya*, which was made available by Professor Bahulkar, and Asaṅga's works, which were available in Shantarakshita Library of the Central Institute of Higher Tibetan Studies, Sarnath. I became convinced that more than 50 per cent of the *sūtra*s could be revisited in light of a Buddhist (or occasionally Jaina) background. Instead of writing an essay on Buddhist influences on the *Yogasūtra*, I decided to write on each *sūtra*, identifying any Buddhist/Jaina relevance if it exists. As a result, the present work has assumed the form of a comprehensive, systematic, and relatively new and radical introduction to the *Yogasūtra*.

I was aware while writing that this is an unconventional work and that a traditional orthodox scholar of Patañjali's Yoga system might not appreciate it for its surface value. However, I was fortunate enough to attract the attention of Professor Maithrimurthi of Heidelberg University, a thorough scholar of Buddhist Abhidharma with an interest in the connection between Yoga and Buddhism, who kindly reviewed my work and made appreciative as well as critical remarks. His comments were a source of encouragement and also helped me both to correct many mistakes and to improve the quality of my work. I am deeply grateful to him.

Along with Professor Mahesh Deokar, I also thank Dr Lata Deokar, who readily found different references from Pali and Buddhist texts for me. Professor Bahulkar, who has always been very helpful and kind, went out of his way to provide me with even more ancient and modern references than I required. I am grateful to him as well as to the friends from whom he sometimes collected the requisite references.

I am also grateful to Professor Kanchana Mahadevan, who invited me to Mumbai University on several occasions in order to share my thoughts on Yoga and Buddhism. I owe a great deal to my friend, Professor Mangesh Kulkarni, who was kind enough to go through the manuscript and subedit certain parts of it.

I am grateful to my Korean friend, Dr Kim Goon Hang, and doctoral students Dong Ho Lim and Jin Sung Park, who became instrumental to my fresh reflections on Yoga and Buddhism.

Yoga and Buddhism have been relevant for me both intellectually and as a part of real-life experience. Though I am not inclined to believe in the otherworldly and supernormal claims of these systems, I have tried, though quite unsystematically, to practice and experiment with their mundane empirical aspects; this has helped me in handling many real-life issues with greater confidence. I took some lessons in Yoga through a course offered by a Yoga master, Dr Samprasad Vinod, who taught Yoga postures in the spirit of Pātañjala-yoga. I also attended some courses in Vipassanā meditation given by the late Satyanarayan Goenka, which had a greater impact. I am grateful to both meditation teachers for their contribution to my life. These practical experiences made my intellectual endeavours with regard to Yoga and Buddhism livelier.

On the practical side of preparing the manuscript, I must mention the individuals who came to my rescue when I needed help with computer typing: Mr S. P. Singh of the Publication Department, Central Institute of Higher Tibetan Studies, Sarnath, and Mrs Nutan Chavan of Pali Department, Savitribai Phule Pune University, Pune. I am grateful to both.

INTRODUCTION

Yoga is one of India's most precious contributions to world culture. In a broad sense, yoga is a method of self-regulation aimed at a noble goal. Different paths have been prescribed in the *Gītā* to pursue the noble goal called liberation—the path of action (*karmayoga*), the path of devotion (*bhaktiyoga*), the path of knowledge (*jñānayoga*) and the path of meditation (*dhyānayoga*). All of these can be subsumed under the broad concept of yoga. To use the term "yoga" in a slightly narrower sense, it is a path of self-discipline which focuses on practices such as postures (*āsana*), breath-regulation (*prāṇāyāma*), control over the senses (*pratyāhāra*), meditation (*dhyāna/samādhi*) or austerity (*tapas*). Depending upon the focus on a particular practice of self-discipline or goal of that practice, yoga is diversified. Hence, we come across the terms Haṭhayoga, Layayoga, Mantrayoga, Rājayoga, Kuṇḍalinīyoga, Dhyānayoga and so on. Pātañjala-yoga, probably the only name coined after its author, is supposed in this context to be the most prestigious.

I. Pātañjala-yoga: the popular approach and the need to revise it

Patañjali, who is recognized as the author of the *Yogasūtra*, is so celebrated among practitioners and lovers of Yoga that he is regarded as the founder of the whole Yoga system. But the question that still requires exploration is whether what is taught and practiced in the name of Pātañjala-yoga is the same as the Yoga advocated by Patañjali in *Yogasūtra*. I suggest that the so-called Pātañjala-yoga, as it is understood and practiced today, has deviated significantly from Yoga as advocated by Patañjali. Three deviations are significant in this context:

(1) Pātañjala-yoga is sometimes identified with Haṭhayoga.
(2) Pātañjala-yoga is sometimes identified with a theistic or non-dualistic form of Yoga.

(3) The possibility of influence on Patañjali by heterodox traditions of self-discipline, particularly those of the Buddhist tradition, has been completely neglected.

It is the third deviation with which I am concerned in this work. However, a brief consideration of the first two deviations is also relevant.

(1) Identification of Pātañjala-yoga with Haṭhayoga

Today, at the popular level, the term "Yoga" refers to the practice of yogic postures (*āsanas*) along with exercises in breath-control (*prāṇāyāma*) and a taste of meditation (which generally consists of the recitation of *Om*). The posture and breath exercises taught in these classes in fact constitute another system of yoga, one known as Haṭhayoga. This system is derived from texts such as *Haṭhapradīpikā* (a fifteenth-century work from Svātmārāma) and *Gheraṇḍasaṁhitā* (a seventeenth-century work by an unknown author). These Haṭhayoga texts also discuss frames (*mudrā*) and fixtures (*bandha*), which are special forms of postures and breath-control. The major goals of these practices are removal of diseases, increase in abdominal fire, longevity and so on. This is the popular understanding of Yoga, which is health- or body-oriented. A body-oriented approach to Yoga has its advantages, and that is why it is very popular. But in carrying the brand of Patañjali, it becomes misleading.

(2) Identification of Pātañjala-yoga with theistic and non-dualistic forms of yoga

Of course, the previously-mentioned Haṭhayoga texts do also refer to spiritual practices such as *dhyāna* and *samādhi*. In fact, in these texts, the practice of *samādhi* is identified as Rājayoga. Svātmārāma, in his work *Haṭhapradīpikā*, asserts that as Haṭhayoga and Rājayoga are interdependent, they should be practiced together.[1] The question is whether Rājayoga, as introduced in Haṭhayoga texts, can be identified with Pātañjala-yoga simply because *samādhi* is central to both of them. While discussing Rājayoga, Svātmārāma explains *samādhi* as both "oneness between *manas* and *ātman*" and "oneness between *jīvātman* and *paramātman*".[2] Similarly, the kind of meditation included under the term "Rājayoga" by the author of *Gheraṇḍasaṁhitā* refers to the idea that "I am Brahman".[3] It is obvious, on the other hand, that the nature of *samādhi* according to Pātañjala-yoga is identity neither between *manas* and *ātman* nor between *ātman* and *paramātman*. In spite of such basic differences, we find in modern times that Pātañjala-yoga is often identified as Rājayoga.

It is true that the concept of *īśvara* has an important role to play in Patañjali's *Yogasūtra*. While accepting the Sāṅkhya metaphysical framework, Patañjali slightly extended it by introducing the concept of ideal *puruṣa*, which could be made into an object of meditation. Meditation on ideal *puruṣa* in the framework of the *Yogasūtra* was supposed to have some special advantages. Patañjali referred to this ideal *puruṣa* by the name *īśvara*. This ideal *puruṣa* is described as omniscient (being pure consciousness) but not omnipotent. Though he is called *īśvara*, he is not supposed to be the creator of the world. The basic Sāṅkhya theory, according to which primordial Nature is the cause of the world, is not affected by the existence of *īśvara*. Since this *īśvara* is free from all actions and passions, he cannot be an object of devotion. He is also called the teacher of teachers. *Īśvara*, untouched by actions and passions, cannot be a teacher in the sense of an instructor, but probably can be a teacher in the sense of an exemplar, that is, the ideal example to be followed or imitated.

However, although *īśvara* was not God according to Patañjali, he was made into one by the commentators. Vyāsa's commentary on *Yogasūtra*, which is supposed to be the oldest commentary, interpreted *īśvara* as a religious object, an object of devotion, to whom an aspirant is supposed to surrender all his actions.[4] This gave impetus to the later commentators to apply more religious or even non-dualist metaphysical colour to the image of Patañjali's *īśvara*. A brief account of how this happened is relevant.

After *Yogabhāṣya*, the period of which could be fourth century or thereafter, the next important commentator of *Yogasūtra* was Vācaspatimiśra (eighth/ninth century CE). In his commentary, *Tattvavaiśāradī*, he argued that the *īśvara* is the creator and destroyer of the world and that there is a beginningless series of the creation and the dissolution of the world.[5]

Vācaspatimiśra was followed by Bhoja (eleventh century CE), the author of *Rājamārtaṇḍavṛtti*, who derived *īśvara*'s comprehensive power from the etymology of the word *īśvara* itself.[6]

The next commentary was *Yogasūtrabhāṣyavivaraṇa* of Śaṅkara. According to some scholars, this Śaṅkara was the same as the well-known Ādiśaṅkarācārya, in which case this commentary would stem from prior to that of Vācaspatimiśra. According to Rukmani, however, this Śaṅkara was different, from between the twelfth and fifteenth centuries CE.[7] Whether he was the same Śaṅkara or not, it is evident in his commentary on the *Yogasūtra* that he does not try to advocate for the metaphysics of non-dualism. What is important for our purpose is that he interprets Patañjali's *īśvara* as God, the omniscient as well as omnipotent (*sarvaśaktimān*) being who produces, maintains and destroys the world. The commentator also provides a detailed argument for the existence of God.[8]

The next important commentary in this series is the fifteenth-century CE *Yogavārtika* by Vijñānabhikṣu. He interprets Patañjali's *īśvara* as f*paramātmā*, the supreme soul or God, the creator and destroyer of the universe, supporting his claim with many quotations from the *Gītā* and *Purāṇas*.[9]

The final commentary in this list is the eighteenth-century *Nāgojībhaṭṭīyavṛtti*, in which an attempt is made to interpret the *Yogasūtra* as a non-dualist Vedānta text. According to Nagojībhaṭṭa, *īśvarapraṇidhāna* means concentration on *īśvara*, which is the same as *parabrahma*, through recitation of *Om*.[10]

Through this commentarial development of Pātañjala-yoga, its original mind-centric nature blurred and it began to be presented as a God-centric or a non-dualistic system of self-discipline. In the modern period, we also see this in Swami Vivekananda's interpretation of Pātañjala-yoga. Vivekananda titled his work on Pātañjala-yoga as *Rajayoga*. As we have seen, the term "Rājayoga" is used in Haṭhayoga texts to connote meditation on *ātman* and *paramātman* or *Brahman* and the identification of the mind with them. Vivekananda may not have been particular about Rājayoga's technical meaning, instead using the term for Pātañjala-yoga mainly to distinguish it from Haṭhayoga. But the fact remains that he was looking at Pātañjala-yoga from a synthetic perspective which combines the atheistic form of Yoga with those of theistic and Advaitic forms.[1] This is clearly indicated by the eighth chapter of *Rajayoga*, "Rajayoga in brief", which includes an account of Yoga from *Kūrmapurāṇa*. The account includes different forms of Yoga. One is Abhāvayoga, where one's self is meditated upon as zero. The other is Mahāyoga, defined as that in which one sees the self as full of bliss, bereft of all impurities and one with God. It also refers to the yoga of eight limbs (verbally the same as those of Patañjali but defined differently), where *dhyāna* is understood as identification with God.

So, in *Rajayoga*, Vivekananda presents Pātañjala-yoga as a Sāṅkhya-based Yoga,[11] on the one hand, but synthesizes it, on the other, with Rājayoga based on theism and Advaitism.[12]

In the twentieth century, there is a growing tendency to mix all three yogas together: Haṭhayoga, which is body-centric; Pātañjala-yoga, which is mind-centric; and Rājayoga, which is God-Brahman-centric. In this synthetic Yoga, the God-Brahman-centric approach dominates the scene. The late B. K. S. Iyengar, a reputed Yoga master, started his class on Haṭhayogic postures with a prayer from Patañjali, indicating that Patañjali was the founder of posture-based Yoga. In the Introduction to *Light on Yoga*, Iyengar is seen as synthesizing yoga as understood in the *Gītā*, *Upaniṣads*, *Haṭhayoga* texts and the *Yogasūtra* of Patañjali. His own conception of yoga, as such, was theistic–non-dualistic, which he seems to have superimposed on Indian thought as a whole.[13]

(3) The neglect of the influence of Buddhism on Pātañjala-yoga

The popular understanding of Yoga completely neglects the influence of Buddhism on Pātañjala-yoga. Many obscure and enigmatic terms, concepts and doctrines mentioned and used by Patañjali can be understood with greater clarity if their Buddhist background is taken into account. Many points made in Vyāsa's commentary can also be understood better against a Buddhist background. In fact, this is one of the main arguments of the present work, one that will be substantiated in due course. However, in order to appreciate the argument, it is necessary to set aside many popular biases about Pātañjala-yoga and look at it from a more historical and rational perspective. Fortunately, there are modern scholars who have taken steps in this direction. In what follows, I will try to outline such a perspective by taking clues from what modern scholars have already done. I aim to show that such a rational and historical perspective will naturally lead to acknowledgement and appreciation of the role of Buddhism in shaping Pātañjala-yoga.

II. Re-visiting Pātañjala-yoga: towards appreciation of the Buddhist roots

Let me begin with some preliminaries. Pātañjala-yoga is a system of yoga contained in the *Yogasūtra*, an aphorismic text which consists of 195 aphorisms (*sūtra*) divided into four chapters: (1) *Samādhipāda* (51 aphorisms); (2) *Sādhanapāda* (55 aphorisms); (3) *Vibhūtipāda* (55 aphorisms); and (4) *Kaivalyapāda* (34 aphorisms).[14] The first chapter focuses on the nature of the goal of meditative practice, namely meditative absorption or deep meditative concentration (*samādhi*), and the different means by which to achieve it. The second chapter focuses on the means aspect (*sādhana*) more systematically. It presents the path of yoga in the form of "yoga of action" (*kriyāyoga*) and "yoga having eight limbs" (*aṣṭāṅgayoga*) and situates this path in the general framework of the four conceptual foundations: the problem of suffering, its cause, its removal and the means to removal (*heya*, *heyahetu*, *hāna* and *hānopāya*). It also explains the cause of the problem in terms of mental defilements (*kleśa*). The third chapter focuses on the supernormal achievements (*vibhūti*) the yoga practitioner can attain with the maturation of his meditative practice. The last chapter mainly deals with the journey of the yogin towards final emancipation, but it also engages in a philosophical debate with idealism. In addition to these main topics, many other topics are discussed by way of stating the advantages of certain practices, clarifying relevant concepts or raising philosophical issues.

The author of the Yogasūtra*: issues regarding his identity and periodicity*

There are many issues around Patañjali, who is popularly known as the author of the *Yogasūtra*. As some modern scholars have pointed out, while Patañjali is found as the name of the *Yogasūtra*'s author in some later commentarial works, it is not found in the *Yogasūtra* itself or in its earliest commentary, *Yogabhāṣya*.[15] Similarly, the name of the *Yogabhāṣya*'s author, said to be Vyāsa or Vedavyāsa, is not found in the *Yogabhāṣya* itself, though it is found in some later commentarial works. Hence, some scholars, like Johannes Bronkhorst and Philipp Mass, hold that the original names of the authors of the *Yogasūtra* and *Yogabhāṣya* may not have been Patañjali and Vyāsa respectively. Though I do not want to rule out this possibility, I use the names Patañjali and Vyāsa in my work to refer to the two authors as a matter of convenience. So, by "Patañjali", I here mean "the author of *Yogasūtra*, who is traditionally named as Patañjali". Similarly, by "Vyāsa", I mean "the author of *Yogabhāṣya* who is popularly known as Vyāsa".

We can notice two major trends among Yoga scholars with regard to the identity and periodization of the *Yogasūtra*'s author: traditional and modern.

Traditional view

Traditional scholars of Indian philosophy are inclined to understand the identity and period of the author of the *Yogasūtra* more on mythological grounds than historical ones. There are two main views under this category. (1) According to one view, Vyāsa, the *Yogasūtra* commentator, is the same as the author of the *Mahābhārata* and thus belonged to the period at the end of the *Dvāpāra-yuga* and the beginning of the *Kali-yuga* (approximately 3000 BCE). Naturally, Patañjali belonged to the same period. This view rests on the doubtful identity of the two Vyāsas: the author of the *Mahābhārata* and the commentator of the *Yogasūtra*. (2) The other view rests on another doubtful identity, that of the three "Patañjalis": the authors of *Vyākaraṇamahābhāṣya*, *Yogasūtra* and *Carakasaṁhitā*. At least two verses are quoted in this context:

> I bow down with folded hands before Patañjali,
> who removed the dirt of mind through yoga,
> dirt of speech through grammar
> and the dirt of body through medical science.[16]

> I salute the king of serpents,
> who destroyed the defects of mind, speech and body
> through *Pātañjala(-Yogasūtra)*, *(Vyākaraṇa-)Mahābhāṣya*
> and the work edited by Caraka (respectively).[17]

According to this view, the period of Patañjali would be the same as that of *Vyākaraṇamahābhāṣya* and *Carakasaṁhitā*, that is, about the second century BCE. However, the statements which assert the identity of the three authors are not found before the tenth century. They occur later, and hence it is more likely that these statements are based on wishful thinking or dogmatic belief than historical evidence.

Though traditional scholars may differ on the period to which Patañjali belongs, they share some other descriptions in common. Accordingly, Patañjali is regarded as the incarnation of Śeṣa, the king of serpents, believed to be the seat of the lord Viṣṇu. In this way, the system of Yoga contained in the *Yogasūtra* is supposed to have divine and unquestionable status.[18] Similarly, traditional scholars commonly refer to the ancient sage Hiraṇyagarbha as the author of *Yogaśāstra*, which formed the basis of Patañjali's *Yogasūtra*.[19] But though the name Hiraṇyagarbha is found in the traditional account, details of his Yoga are not. So, the character of Hiraṇyagarbha could be mythological rather than historical. Whatever be the case, the traditional view in general holds that Patañjali's Yoga system has a divine origin and a Vedic background. Hence, the question of heterodox tradition having any influence on Patañjali's Yoga is ruled out.

Modern view

The modern view does not regard Patañjali as the incarnation of Śeṣa, nor does it stretch its period to the mythological period of Vedavyāsa (the author of *Mahābhārata*). It considers Patañjali as a historical personality rather than a mythological figure. Again, we have two views under this category.

(1) Second century BCE as the period of Patañjali: the identity of two Patañjalis

This view accepts the identity of the two Patañjalis: the Patañjali of *Yogasūtra* and the Patañjali of *Vyākaraṇamahābhāṣya*. Dasgupta (1975: 230–233) argues in favour of this view in *History of Indian Philosophy*. According to Dasgupta, though there is no sufficiently strong evidence in favour of the identity of the two Patañjalis, there also is nothing which could be considered as conclusively against it. According to this view, the period of the author of the *Yogasūtra* would be second century BCE. Dasgupta (1975: 229) also holds that Patañjali was not the founder of the Yoga system but rather

> collected the different forms of Yoga practices, and gleaned the diverse ideas which were or could be associated with the Yoga, but grafted them all on the Sāṅkhya metaphysics, and gave them the form in which they have been handed down to us.

The diverse ideas gleaned by Patañjali include Buddhist ideas such as *saṁsāra*, *avidyā* and *abhiniveśa*, as well as the four sacred truths of Buddhism.

However, Dasgupta's thesis that the author of the *Yogasūtra* belongs to the second century BCE fails to explain the criticism of Buddhist idealism contained in its fourth chapter. To address this difficulty, Dasgupta maintains that the *Yogasūtra*'s fourth chapter must be a later interpolation.

There are in fact two main difficulties with the thesis that the *Yogasūtra*'s author belongs to the second century BCE. First is that the idea of the identity of two Patañjalis is not sufficiently justified. As Woods (2003: xv–xvii) holds, the identification of the two Patañjalis is not confirmed by a comparison of their philosophical concepts. It can be observed in this context that the two Patañjalis do not seem to have the same perspective on language. The grammarian Patañjali is known for holding the *sphoṭa* theory of language, according to which words as meaning-bearing units are fixed entities that only become manifest through sounds. The Patañjali of the *Yogasūtra*, on the other hand, suggests that words, objects and cognitions are often mixed up in our language due to mutual superimposition, and that a yogin needs to focus on them as separate entities (YS III.17). He also maintains that many words in our language simply do not stand for anything in the world. They give rise to mental constructions (*vikalpa*) which have no reference to reality (YS I.9). This perspective on language seems very different from that of the grammarian Patañjali, who looked at language as a well-organized system.

The second difficulty is that the whole of the *Yogasūtra* (with all four chapters) cannot be maintained as a single text. This was in fact Dasgupta's hypothesis, which caused him to divide the text into two parts: the first three chapters as the original *Yogasūtra* and the fourth chapter as a later addition. However, there is no basis for such a division of the text except the thesis regarding the identity of the two Patañjalis. The argument is circular. This brings us to the second modern view.

(2) Second half of the fourth century CE as the period of Patañjali

The second modern view questions the alleged identity of the two Patañjalis. Once the thesis regarding the identity of the two Patañjalis is rejected, there remains no need to divide the *Yogasūtra* into the first three chapters (as authored by Patañjali) and the final one (a later addition). The whole text can belong to some later period and be authored by a single author.

In this context, there are modern scholars who point out that some aphorisms in the *Yogasūtra*'s third and fourth chapters (YS III.11–15, YS IV.12) contain a response to Vasubandhu's discussion of Sarvāstivāda (the view that things exist in all three times), while some aphorisms in the fourth chapter (YS IV.14–21) contain criticism of the idealist Buddhist doctrine of *vijñaptimātratā* ("consciousness-only").[20] Hence, they consider the *Yogasūtra* to be contemporaneous with Vasubandhu's works. Accordingly, Patañjali's period is ascertained as the second half of the fourth century CE or, rather, between 325 and 425 CE.[21] In this work, I follow this view of modern scholars.

The relation between Yogasūtra *and* Yogabhāṣya

Though the basic text of Patañjali's Yoga system is the *Yogasūtra*, the text is generally studied alongside its earliest commentary, which is attributed to Vyāsa. Sometimes the two texts together are treated as one uniform unit: *Pātañjala-yogaśāstra*.[22] The dating of Vyāsa and his relationship with Patañjali become relevant issues in this context. Vyāsa's period could range from 350 to 850 CE, so, how close Vyāsa must have been to Patañjali is a question. Some modern scholars date Vyāsa as very close to the author of the *Sūtras*. Bronkhorst (1985) argues that the author of the *Yogabhāṣya*—who was either named Patañjali or wrote *Yogabhāṣya* under the name of Patañjali—was himself the compiler of the *Yogasūtra*.

In order to substantiate his claim, Bronkhorst offers an interesting piece of "evidence".[23] In the commentary for YS II.27, the Bhāṣyakāra interprets *tasya* as "*pratyuditakhyāteḥ*". The term "*pratyuditakhyāti*" does not occur in the *Yogasūtra* anywhere, but it does occur in *Yogabhāṣya* I.16. From this, Bronkhorst infers that the Sūtrakāra is here referring to the word *pratyuditakhyāti* used in the Bhāṣya.

I differ from Bronkhorst on this point. His argument reads the concerned Sūtra and Bhāṣya only at surface level. "*Pratyuditakhyāti*" is not a special technical term which the Sūtrakāra should have used before if he wanted to talk about a person in whom *khyāti* has arisen in the aphorism YS II.27. *Khyāti* and *vivekakhyāti* are certainly important technical terms. So, when the Bhāṣyakāra interprets *tasya* as *pratyuditakhyāteḥ*, he is referring back to *khyāti*, that is, *vivekakhyāti*, and its use (the term *vivekakhyāti*) by the Sūtrakāra in the immediately preceding aphorism. So, in the Bhāṣya statement, namely, "*tasyeti pratyuditakhyāteḥ pratyāmnāyaḥ*", the expression *tasya* refers to "the person in whom *khyāti* has arisen". It does not refer to "the expression, namely *pratyuditakhyāti*"; and this solves the problem.

So, I do not find any strong evidence to support the view that the Bhāṣyakāra himself must have been the Sūtrakāra for at least some *sūtras*. It is consistent to presume that the Sūtrakāra and the Bhāṣyakāra were different personalities. As I have shown in my explanations of some of the aphorisms and Bhāṣyakāra's interpretation of them, many discrepancies can be found if the aphorisms are read independently of any commentary or in light of the Buddhist background.[24]

Hence, although I may grant the view accepted by modern scholars such as James Woods, Gerald Larson (as in Larson and Bhattacharya 2008: 53) and others that the period of Vasubandhu's *Abhidharmakośa* (with *bhāṣya*) and the *Yogasūtra* is broadly the same (namely 350–400 CE, or 325–425 CE as Philipp Mass (2013: 65) accepts), I do not accept the advocacy of complete hermeneutic unity between *Yogasūtra* and *Yogabhāṣya*. I feel there must have been a considerable gap—an ideological and hermeneutical gap at least, and perhaps a chronological gap as well—between them. Neither do I think that the Bhāṣyakāra himself was the Sūtrakāra or that the former was in close contact with the latter such that the former understood the latter's intention correctly or both conceptualized yoga in the same way.[25]

The hermeneutic gap between Yogasūtra *and* Yogabhāṣya *and the role of Buddhist influence*

That the Bhāṣyakāra understood the intentions of the Sūtrakāra correctly, and that Yogabhāṣya is an authentic interpretation of the *Yogasūtra*, is a big assumption, albeit one made by most of the traditional and modern commentators. A scholar like Bronkhorst may doubt this assumption, but does not do so uniformly. For example, in his long argument about the set of Yoga aphorisms of I.30–40, Bronkhorst accepts the Bhāṣya explanation of the expression "*viṣayavatī pravṛttiḥ*" in YS I.35 as "consciousness of supernormal odour, taste, colour, touch and sound", though this does not follow from the aphorism.[26] However, he interprets the compound word "*sthitinibandhanī*" as a *Bahuvrīhi* compound by going against the explicit Bhāṣya interpretation of it as a Tatpuruṣa compound.[27] In spite of such occasional deviations from the *Yogabhāṣya* interpretations, Bronkhorst seems to believe that the Bhāṣyakāra must have known what the Sūtrakāra meant and that, by and large, he interpreted the Sūtrakāra correctly.

But Bronkhorst has made other interesting claims about the Bhāṣyakāra that run contrary to his general beliefs. For instance, he claims it likely that the skills of the Bhāṣyakāra were primarily or even exclusively theoretical and that he may not have had any direct experience of yogic states (Bronkhorst, 1984). Elsewhere (Bronkhorst, 1981), he observes that *Yogabhāṣya* was never meant to be representative of anything but the Sāṅkhya philosophy. A similar approach can be found in the writings of Georg Feuerstein and J. W. Hauer. Feuerstein (1979: 25) quotes Hauer approvingly, "The commentators subsequent to Vyāsa, even already Vyāsa himself, instead of presenting the genuine philosophy of Yoga, often foist on Yoga the philosophy of Sāṅkhya. For this reason, they are to be used with caution".

I would like to juxtapose these remarks about the *Yogabhāṣya* and its author with remarks about the Yoga in the *Yogasūtra* made by scholars such as Erich Frauwallner and Gerald Larson. Frauwallner understood Sāṅkhya as having two directions. One was the philosophical school led by Vārṣagaṇya; the other was Yoga, which was the practical way. He referred to the two directions as the philosophical school and the Yoga school respectively. He makes the following statement about the two schools:

> While the Philosophical School is most of all indebted to the most important philosophical system which had developed at that time besides the Sāṅkhya, the Vaiśeṣika, the Yoga school borrowed its important new ideas on which the doctrine of deliverance is built and established from a different source which was no other than Buddhism.
>
> *(Frauwallner, 1984: 323)*

Scholars such as de La Vallée-Poussin have identified many locations in the *Yogasūtra* (and also *Yogabhāṣya*) where Buddhist influence is undeniable. In light of these researches, Gerald Larson and Bhattacharya (2008: 44) holds that the *Yogasūtra* "represents a hybrid formulation, a conflation of old Sāṅkhya philosophy, and an attempt to codify old meditation traditions coming from the last two centuries BCE and the first centuries CE".[28] In a paper published in 1989, Larson states his conjecture as to how such a hybrid system emerged:

> There were two streams of early systematic philosophizing in India, namely the *Ṣaṣṭitantra* of Sāṅkhya and the *Abhidharma* of Sarvāstivāda and Sautrāntika (and possibly other Buddhist traditions as well). In the first centuries of the Common Era or at about the time of Vārṣagaṇya these two traditions began to interact with each other and systematic Yoga philosophy is a hybrid form of that interaction probably occasioned by the *Abhidharmakośa* and *Bhāṣya* of Vasubandhu.
>
> *(Larson, 1989: 134)*

Larson rightly leaves room for "other Buddhist traditions". I will fill this blank space with the early Yogācāra Buddhism developed by Asaṅga. One could say that Yogācāra must have contributed to Pātañjala-yoga through a dialectical process. The point can be explained as follows.

In some *Upaniṣads* and in the *Mahābhārata* (*Mokṣadharma* and *Bhagavadgītā*), we find references to Yoga (which accepted either Sāṅkhya or Upaniṣadic metaphysics). Against this, some Buddhists tried to develop an alternative system of Yoga which accepted neither God nor an eternal self. Asaṅga was a representative of this Buddhist antithesis and posed a challenge to Brahmanical yoga by developing a system, Yogācāra, through his voluminous *Yogācārabhūmi*. Through this work, he also attacked different Sāṅkhya and Upaniṣadic doctrines.[29] It is possible to argue that Patañjali's *Yogasūtra* was a response to the Buddhism-based Yoga system constructed by Asaṅga. Patañjali incorporated many Buddhist elements in his response but gave supremacy to the metaphysics of Sāṅkhya.

Here I do not mean to say that the "Buddhist theory" of meditation was completely different and isolated from the "Brahmanical form" of Yoga. In fact, it is an interesting area of study in terms of how Buddhist meditation is linked to Brahmanical conceptions of meditation, and indeed is rather a two-way relation. The *bodhisattva* Gautama learnt the traditional forms of meditation and austere practices available in his time, experimented with them, discarded some after experimentation, modified and refined some of them and also introduced new forms by transcending them. In this way, Buddhist meditation stems from the ancient forms of meditational and austere practices. However, the Brahmanical meditational practices from which Buddhist meditation stems were available in later times only in the form in which they were recorded and preserved in the Buddhist texts. In fact, they formed a part of Buddhist meditation in a broad sense of the term. It was this Buddhist theory of meditation, in its broad sense, which was the background of *Yogasūtra*, and this Buddhist background was influential in shaping the Yoga system of Patañjali.

If Buddhism plays such a dominant role in shaping Patañjali's Yoga system, then it would be legitimate to expect a judicious commentary on the *Yogasūtra* to interpret the text by throwing light on all influencing factors as background factors. Unfortunately, *Yogabhāṣya* does not fulfil this expectation. The Bhāṣyakāra must have been aware of this Buddhist background, as is clear from his own use of many Buddhist terms and ideas. But he seems to have knowingly avoided acknowledging the Buddhist influence. He interprets the whole of the *Yogasūtra* as a Sāṅkhya text. He acknowledges the Buddhist position only when it is to be criticized or when the Sāṅkhya position is to be defended against it.

Unfortunately, the Bhāṣyakāra's narrow interpretation of *Yogasūtra* came to be treated as authentic at the hands of the later commentators, and the Buddhist background got completely side-tracked. This also became a constraining factor in modern scholarship on Pātañjala-yoga. Hence, modern scholarship has largely rotated around the Bhāṣya-centric understanding of the *Yogasūtra*.

In order to break the circle and arrive at a comprehensive understanding of the *Yogasūtra*'s thought content, an important step is to clarify the Buddhist background of *Yogasūtra*. Though scholars like de La Vallée-Poussin have undertaken some foundational work in this regard, it is necessary to identify the Buddhist sources more extensively and systematically in terms of terminology, concepts and doctrines. This is one of the driving forces behind this book.

The Buddhist background literature

While investigating the Buddhist roots of Pātañjala-yoga, one question concerned the choice of the Buddhist literature. Many scholars agree that the authors of *Yogasūtra* and *Yogabhāṣya* were exposed to Vasubandhu's *Abhidharmakośa* and *Bhāṣya*. I thought that the two authors might also have been exposed to the Abhidharma literature and Yogācāra literature of Asaṅga. In fact, Asaṅga's *Yogācārabhūmi* covered many aspects of Buddhism, including Śrāvakayāna and Bodhisattvayāna. Hence, many terms and concepts in *Yogasūtra* and *Yogabhāṣya* can be traced to Asaṅga's works. I have referred to Pali literature rarely because I thought that the *Yogasūtra* and *Yogabhāṣya* authors may not have been exposed to early Buddhism through Sanskrit works like *Abhidharmakośabhāṣya* and *Śrāvakabhūmi* rather than directly. But sometimes I had to make an exception. In the early Buddhist theory of meditation, *vitakka* (Sanskrit: *vitarka*) and *vicāra* were conceptualized differently from in Sanskrit Abhidharma literature. I thought that both conceptions were reflected in Patañjali's theory of meditation. Hence, along with *vitarka* and *vicāra* of the Sanskrit Abhidharma, I have referred to Pali Buddhist conceptions of *vitakka* and *vicāra* as a background for Patañjali's theory of meditation.[30]

In my survey, I found that more than 50 per cent of the *Yogasūtra*'s aphorisms can be said to have a background in Buddhism, directly or indirectly. Less than 40 per cent of the aphorisms can be said to have a background in Sāṅkhya or other orthodox sources, directly or indirectly. Some aphorisms have both backgrounds, while others do not have either background specifically. I decided to write about each aphorism separately in order to offer a clearer picture of the background to the *Yogasūtra* (and *Yogabhāṣya*). Hence, the book has been written in the form of an aphorism-wise gloss on *Yogasūtra*. After providing the citation, transcription and translation of each aphorism, I comment on its possible source: Buddhist, Jaina or Sāṅkhya. In the case of Buddhist sources, I comment in greater detail, and also point out on many occasions the Buddhist background of Vyāsa's commentary.

In this book, I try to show the relevance of the Abhidharma literature of Sarvāstivāda, Vaibhāṣika and Sautrāntika Buddhism, as well as Asaṅga's Yogācāra writings, for understanding the *Yogasūtra*'s many aphorisms in addition to terms and passages in the *Yogabhāṣya*.

The objective of the present study

The objective of my work is neither to refute nor to defend Patañjali. Most students of *Yogasūtra* start with the assumption that Patañjali has given an authentic and coherent system of Yoga and that our job is to understand how it is coherent and authentic. I do not start with this assumption. Patañjali's system of Yoga likely is not a coherent system because it is an attempt to synthesize two soteriological systems—Sāṅkhya and Buddhism—which are mutually inconsistent in many ways. Traditional and modern commentators and scholars have both emphasized how the orthodox philosophical tradition in general and Sāṅkhya in particular have shaped the Yoga system of Patañjali. A few scholars have indicated that Buddhism must have also shaped Patañjali's Yoga system, but they have mainly indicated the direction in which it is possible to proceed. My attempt in this work is to advance in the direction shown by this handful of scholars. Hence, I have tried to investigate the Buddhist background of Patañjali's Yoga as extensively as possible. I have found that many Yoga aphorisms make different sense than traditionally understood, and sometimes better sense, if read in the light of Buddhist literature. Occasionally I found that Vyāsa's interpretation of some aphorisms went wrong because he failed to take the Buddhist background of those aphorisms into account. This is not to say that Vyāsa's commentary itself is not influenced by Buddhism. Though he interprets *Yogasūtra* as a Sāṅkhya text, many of the ideas he uses in the course of his interpretation are influenced by Buddhism. The aim of my work is to bring to the foreground the Buddhist context of both *Yogasūtra* and *Yogabhāṣya*.

Naturally, what the work expects from its readers is different from what orthodox commentaries and monographs on the *Yogasūtra* expect from their readers. Most of the works on Pātañjala-yoga presuppose the orthodox framework of beliefs. They project Patañjali's Yoga as making its contribution against the background of a Vedic, Upaniṣadic and Sāṅkhya metaphysical and soteriological framework. They neglect the possibility of a heterodox philosophical tradition influencing the Yoga of Patañjali. Moreover, they presuppose that *Yogasūtra* is a fully consistent and authoritative text and try to interpret it in such a way that the consistent and authoritative character of the text is preserved.

The present work does not presuppose the *Yogasūtra* to be a consistent and authoritative text. Nor does it presuppose that the text strictly follows the orthodox metaphysical and soteriological framework. It rather makes an appeal to appreciate the role of heterodox tradition, particularly that of Buddhism, in shaping the Yoga system of Patañjali. The work treats the *Yogasūtra* as an attempted but not necessarily successful synthesis of Sāṅkhya and Buddhism.

By introducing the *Yogasūtra* anew, this work makes an appeal for an appreciation of the Buddhist roots of Pātañjala-yoga. It also involves a claim that neglect of the Buddhist influence on Pātañjala-yoga at the hands of traditional and modern scholars has done injustice to Buddhism and to the proper understanding of Pātañjala-yoga. And though a reader with an orthodox inclination may not accept such an extreme claim, he or she can still find this work useful, as many terms, concepts and doctrines in the *Yogasūtra* and *Yogabhāṣya* can be understood with greater clarity and richness in light of their counterparts in Buddhism, as suggested in the comments on the aphorisms in this work. This is the minimum claim of this work.

A summary of the work

The first four chapters of the book consist of interpretation and discussion of the four quarters (*pāda*) of the *Yogasūtra*. The fifth chapter consists of concluding observations and comments.

The first chapter discusses the first quarter of the *Yogasūtra*, *Samādhipāda*, which defines and discusses the concept of meditative absorption (*samādhi*) in its various aspects. The aphorisms here classify meditative absorption and meditative attainment in terms of gross and subtle objects, possessing and not possessing object-consciousness and containing and not containing the seed of latent impressions. They discuss obstacles on the way to meditative absorption and the different objects of concentration prescribed for removing the obstacles. The chapter translates Patañjali's aphorisms and comments on them. The comments reveal how the terms and concepts discussed by Patañjali—such as *samādhi*, *samāpatti*, *vitarka*, *vicāra*, *vikalpa*, *abhyāsa*, *vairāgya*, the five means (*upāya*), the four *bhāvanās*, *praṇidhāna*, three-fold *prajñā* and many others—are understood well against the Buddhist background.

The second chapter, *Sādhanapāda*, discusses the second quarter of the *Yogasūtra* and deals with the means to the final goal, namely isolation (*kaivalya*), or the penultimate goal, which is called seedless meditative absorption (*nirbīja-samādhi*). The aphorisms in this quarter present the Yoga of action (*kriyāyoga*) and also the eight-limbed Yoga. They present the framework of the four foundational concepts: what it is to be abandoned (namely, the suffering yet to come); the cause of this; abandonment of this; and the means to abandonment. The comments in this chapter bring forth the background to Patañjali's aphorisms, whether Sāṅkhya, Buddhist or Jaina. They bring out the Buddhist background of Patañjali's notions, such as *kleśa*, *avidyā*, *asmitā*, *rāga*, *dveṣa* and *abhiniveśa*; four foundational concepts related to suffering and its abandonment; seven-fold ultimate wisdom (*saptadhā prāntabhūmiḥ prajñā*); posture; and breath-regulation. They also bring out the Jaina background to Patañjali's concept of *yama* as the great vow.

The third chapter discusses the third quarter of the *Yogasūtra*, *Vibhūtipāda*, which elaborates on different supernormal powers that a yogin can attain, according to Patañjali, by using the technique of integrated concentration (*saṁyama*). The aphorisms in this quarter begin with definitions of the last three limbs of the eight-limbed yoga, namely fixed attention (*dhāraṇā*), meditation (*dhyāna*) and meditative absorption (*samādhi*). Then, the notion of integrated concentration (*saṁyama*)—a key notion of the Patañjali's theory of supernormal knowledges and powers—is defined. They include physical as well as cognitive powers, worldly as well as spiritual powers. All such powers are claimed to be the result of integrated concentration on relevant objects. Patañjali, however, warns here that though supernormal powers are achievements in the extravert state of mind, they are obstacles in meditative absorption. The comments on Patañjali's aphorisms reveal how his claims about supernormal knowledges and powers largely reflect claims about *abhijñā*s and *ṛddhi*s made by the Buddhist tradition as represented by Asaṅga and Vasubandhu. The chapter also brings out the similarity between Vasubandhu's and Vyāsa's views about the nature of the cosmos and the take of Patañjali and Vyāsa on the Sarvāstivāda school of Buddhism.

The fourth chapter discusses the fourth and final quarter of the *Yogasūtra*, *Kaivalyapāda*, which deals with the journey of the yogin towards final emancipation, that is, *kaivalya*. Patañjali also engages here in philosophical debate with the Buddhist doctrines of momentariness and idealism. The other topics dealt with are different sources of supernormal powers; creation of mind as a supernormal power; and the doctrine of karma and its fruition. The comments on these aphorisms reveal how the Buddhist background—consisting of the theory of supernormal powers, the doctrines of momentariness, idealism and the Mahāyāna doctrine of ten planes (*bhūmi*), with *dharmameghā* as the highest plane—are relevant for understanding Patañjali's claims.

The fifth and concluding chapter argues that through his theory of Yoga, Patañjali tries to synthesize two stories of suffering and emancipation: the empirical–psychological story rooted largely in Buddhism and the story based on Sāṅkhya metaphysics. The chapter explains how Patañjali integrates the two stories. It argues that although Patañjali adopts many inputs from the Buddhist story of suffering and emancipation, he attempts to situate these in the realist and eternalist framework of Sāṅkhya. The chapter also points out how Vyāsa's commentary many times deviates from meanings of the aphorisms which appear to be more natural and consistent.

The concluding chapter is followed by three appendices. Appendix I deals with the Buddhist theory of Yoga and meditation, while Appendix II provides an explanation of Asaṅga's account of supernormal powers. These appendices aim to equip the reader to understand Patañjali's Yoga in a better comparative light. Appendix III offers a list of points on which Vyāsa seems differ from Patañjali's aphorisms.

Studying Pātañjala-yoga in the light of Buddhism has enriched my understanding of both. I hope readers interested in Yoga and Buddhism will have a similar experience while reading this work.

Notes

1 "*haṭhādvinā rājayogo rājayogaṁ vinā haṭhaḥ* | *na sidhyati tato yugmam āniṣpatteḥ samabhyaset* ||" (HP II.76). [Rājayoga does not become successful without Haṭhayoga and the latter does not become successful without the former. Therefore, one should practice them jointly till the fulfilment.]

2 "*salile saindhavaṁ yadvat sāmyaṁ bhajati yogataḥ* |
Tathātmamanasor aikyaṁ samādhir abhidhīyate ||
tatsamaṁ ca dvayor aikyaṁ jīvātmaparamātmanoḥ |
praṇaṭasarvasaṅkalpaḥ samādhiḥ so'bhidhīyate ||" (HP IV.5,7). [Just as the salt gets assimilated with water when mixed with it, likewise the oneness between *manas* and *ātman* (which is brought about by mixing the former with the latter), is called *samādhi*. In a similar way

the oneness between the individual self and the supreme self, is called *samādhi*, the state in which all desires are destroyed.]

3 "*ahaṁ brahma na cānyo'smi brahmaivāhaṁ śokabhāk* | *Saccidānandarūpo'haṁ, nityamuktasvabhāvavān* || " (GS 7.4). [I am Brahman and am no one else. I am only Brahman and not a bearer of suffering. I am of the nature of the real, consciousness and bliss; "permanently liberated" is my nature.]

4 "*praṇidhānād bhaktiviśeṣād āvarjita īśvaras tam anugṛhṇāty abhdhyānamātreṇa*" (YB I.23). [*Praṇidhāna* means a special kind of devotion. *Īśvara*, when propitiated by such a devotion, favours the aspirant on account of simple meditation.] "*īśvarapraṇidhānaṁ sarvakriyāṇāṁ paramagurāv arpaṇaṁ tatphalasaṁnāso vā*" (YB II.1). [*Praṇidhāna* of *īśvara* means surrendering all actions to the greatest teacher or sacrificing the fruits of actions.]

5 "*anādau tu sargasaṁhāraprabandhe sargāntarasamutpannasañjihīrṣāvadhisamaye pūrṇe mayā sattvaprakrṣa upādeya iti praṇidhānaṁ kṛtvā bhagavān jagat sañjahāra*" (TV I.24). [In the beginningless series of creations and dissolutions, when the time limit for completion of the interval after the last creation is up, the lord makes a resolve, "I should cause an excess in *sattvaguṇa*" and he dissolves the world.]

6 "*Īśvara īśanaśīlaḥ icchāmātreṇa sakalajagaduddharaṇakṣamaḥ*" (RMV I.24). ["*Īśvara*" means one who has tendency to rule, one who is capable of uplifting the whole world by his mere wish.]

7 For Rukmani's argument, see the introduction of Rukmani (2001). Harimoto (2004) has questioned the strength of her argument.

8 See YBV I.24–5. About the pervasive power (*sarvaśakti*) of *īśvara*, Śaṁkara says, "*tathā śaktir api sātiśayatvena vivardhamānā kāṣṭhāṁ prāpnoti sa sarvaśaktiḥ. Tena jagannirmāṇa-sthāpanopasaṁhārādi-kartṛtvasiddhiḥ*" (YBV I.25). [Similarly the power, which can be superseded, can go on increasing until it reaches the highest limit. In this way he is omnipotent. By this is established his acts such as creation, sustenance and destruction of the world.]

9 *Yogavārtika* on the aphorisms regarding *īśvara* (YV I.23–29) is replete with such quotations.

10 "*vakṣyamāṇalakṣaṇeśvarasya parabrahmādiśabdavācyasya aupādhikaiśvaryopa-lakṣitasya cinmātrapruṣaviśeṣasya praṇidhānaṁ 'tajjapas tadarthabhāvanam' iti vakṣyamāṇam*" (NBV I.22). [*Īśvara*, which will be defined later, is the one who is denoted by the expressions like "*parabrahman*", who is characterised by lordship, which is conditional, and who, as special kind of spirit, is pure consciousness. "*Praṇidhāna*" means what will be described by the aphorism, "Repetition of that ("Om") means contemplation on the meaning of that".]

11 Hence while commenting upon *Yogasūtra* I.25, Vivekananda says, "The yogins, however, do not mention many ideas about God such as creating, God as creator of the Universe is not meant by the *īshvara* of the yogins".

12 Another iconic personality who practiced and theorized about Yoga in modern times was Aurobindo. He is known for his integral Yoga, which synthesises all three aspects—body-centric, mind-centric and a spirit-God-centric—in one. I do not consider it here because Aurobindo present integral Yoga not as identical with Pātañjala-yoga but rather as his own formula of Yoga.

13 As he says, "In Indian thought everything is permeated by the Supreme Spirit (*Paramātmā* or God) of which the individual human Spirit (*jīvātmā*) is a part. The system of yoga is so called because it teaches the means by which the jīvātmā can be united to or be in communion with the Paramātmā, and so secure liberation" (Iyengar, 1965: 21). This understanding of Yoga is problematic. The conception of yoga he presents is not the common feature of the Indian thought. The concept of yoga necessarily presupposes neither *paramātmā* nor *ātmā*.

14 The arrangement of the previously-mentioned 195 aphorisms is not perfect, but the author seems to have made some adjustments in order to make the four chapters comparable in size. For example, the second chapter ends with the fifth limb of the eight-limbed yoga, while the last three limbs are introduced in the first eight aphorisms of the

third chapter. Similarly, the discussion of supernormal powers spills over into the third chapter and continues through the first six aphorisms of the fourth chapter. But if the chapters are arranged according to the topics dealt with in the aphorisms, then the chapters would contain 51, 63, 53 and 28 aphorisms respectively. This would make the last chapter too short (it is already very short). To minimise the unevenness of the size of the chapters, the author might have divided the chapters the way he has done.

15 In his commentary on YS III.44, Vyāsa refers to Patañjali as holding the view that a substance (*dravya*) is a collection of parts inseparably related to the whole (*ayutasiddha*). This view comes close to the view of the grammarian Patañjali's but does not tally with it fully, as shown by Woods (2003: xv–xvii).

16 "*yogena cittasya padena vācāṁ malaṁ śarīrasya ca vaidyakena | yo'pākarot taṁ pravaraṁ munīnāṁ patañjaliṁ prāñjalir ānato'smi ||*" (quoted by Woods 2003: xiv, from Śivarāma's eighteenth-century text).

17 "*pātañjalamahābhāṣyacarakapratisaṁskṛtaiḥ | manovākkāyadoṣāṇāṁ hantre' dhipataye namaḥ ||*" (quoted by Abhyankar 2006: xvii, from Cakrapāṇidatta's eleventh-century CE commentary on *Carakasaṁhitā*).

18 "*patañjaliś ca sākṣāc cheṣāvatāra iti tatpratipāditeṣu siddhānteṣu na leśato'pi saṁśayasyāvakāśaḥ*" (Abhyankar 2006: XIII)

19 Abhyankar in this context refers to the statement "*hiraṇyagarbho yogasya vaktā nānyaḥ purātanaḥ*" [Hiraṇyagarbha was the ancient author of Yoga and no one else].

20 Woods argues on this line at length and many later scholars generally accept it.

21 Philipp Mass (2013: 65) claims that *Pātañjalayogaśāstra* (which, according to him, is a unified whole of *Yogasūtra* and *Yogabhāṣya*) can be dated with some confidence to between 325 and 425 CE. While this date can be assigned to the *Yogasūtra*, it is doubtful whether the two texts can be regarded as a uniform whole. I express my reservations about this claim in the next section.

22 Philipp Mass (2013: 57) shows that this understanding *Pātañjalayogaśāstra* as a unified whole can be seen in Vṛaṣabhadeva's commentary (circa 650 CE) on *Vākyapadīya* and in many sources dated from the tenth century on. However, this is not sufficiently strong evidence to show that Pātañjala-yoga must have been a unified whole or that *Yogasūtra* and *Yogabhāṣya* had a common author. It is quite possible that Bhāṣyakāra's interpretation, which suppressed the Buddhist influence and depicted *Yogasūtra* as an essentially Sāṅkhya work, was regarded as so authentic and indispensable for understanding the *Yogasūtra* by later commentators that the two texts were treated as a single uniform text.

23 Philipp Mass (2013: 63–5) refers to this evidence.

24 For a collective picture of these discrepancies (as I understand them), see Appendix III.

25 Hence, I do not agree with the conclusion Philipp Maas (2013: 68) draws regarding *Pātañjalayogaśāstra*: "it is advisable from a philosophical and historical point of view to accept, at least hypothetically, that this work is the result of single, roughly datable philosophical authorial intension". On the contrary, I suggest that the authorial intentions of Sūtrakāra and Bhāṣyakāra must have been different. Sūtrakāra's intention was to construct a system of yoga by incorporating empirical–phenomenological elements from Buddhist and similar traditions and the metaphysical framework from Sāṅkhya. Bhāṣyakāra's intention was to interpret the whole system of yoga as a form of Sāṅkhya system.

26 Bronkhorst (1984). I translate the aphorism as, "Or an attitude (of sense organ or mind) arising along with an object, (when concentrated upon), conduces to mental steadiness". See my comments on YS I.35. If we accept this interpretation, which is close to the original aphorism, then Bronkhorst's claim that aphorisms I.35–38 are addressed to advanced practitioners is weakened.

27 "*etā vṛttaya utpannāḥ cittaṁ sthitau nibadhnanti*" (YB I.35).

28 I broadly appreciate Larson's view that Pātañjala-yoga was a hybrid of Sāṅkhya metaphysics and Buddhist meditative tradition. However, the contribution of Jaina soteriology should also be included in this Hybrid.

29 YAB: 118–60 (the section "*Paravāda*: The Opponents' Positions"). The opponents' positions criticised by Asaṅga include (1) effect is present in the cause (*hetuphalasadvāda*); (2) "causation as manifestation" (*abhivyaktivāda*); (3) things exist in the past as well as the future (*atītānāgatadravyasadvāda*); (4) eternalism (*śāśvatavāda*); and (5) God or some such thing is the creator (*īśvarādikartṛvāda)*

30 See Appendix I, Part I.

1

ON MEDITATIVE ABSORPTION (*SAMĀDHIPĀDAḤ*)

In the fifty-one aphorisms of the first chapter, *Samādhipādaḥ*, Patañjali introduces the concept of meditative absorption (*samādhi*) in its various aspects. What follows is an aphorism-wise summary of the first chapter of the *Yogasūtra*.

(1–11): Patañjali defines meditative absorption and situates it in relation to the seer as well as to the epistemological and psychological context of mental modifications or states (*citta-vṛttis*). In his commentary, Vyāsa introduces the doctrine of mental planes (*cittabhūmis*). **(12–16):** Patañjali then states and explains the two major means to meditative absorption: practice and detachment. **(17–22):** He introduces the two types of meditative concentration—with object-consciousness (*samprajñāta*) and without object-consciousness—and explains how they become available to one by birth or through application of means. **(23–29):** Patañjali introduces the notion of supreme being (*īśvara*), resolve regarding the supreme being, meditative concentration on him through utterance of *Om* and the advantages of this method. **(30–31):** He introduces the obstacles on the way to meditative absorption. **(32–40):** Patañjali states various means and objects of concentration prescribed for removing obstacles and achieving peace, tranquillity and stability of mind, through which one can also achieve control over nature. These means and objects include the cultivation (*bhāvanā*) of four sublime attitudes, breath, a mental attitude with its object, special mental attitudes, an attachment-free person, dream, sleep or any object of one's inclination. **(41–46):** Patañjali introduces the nature and types of meditative attainment (*samāpatti*) and identifies them with meditative absorption with seed (*sabīja-samādhi*). **(47–51):** Finally, Patañjali describes the journey of the meditation practitioner from absorption with seed (*sabīja-samādhi*) to absorption without seed (*nirbīja-samādhi*), which is made possible by the truth-bearing wisdom (*ṛtambharā prajñā*) which arises from the practice of the former.

The comments on these aphorisms will reveal how, along with the Sāṅkhya framework, different concepts and doctrines of Buddhist psychology and soteriology must have influenced Patañjali and Vyāsa and how these are relevant for understanding various topics introduced and explained by them in the chapter.

अथ योगानुशासनम् ॥1॥

Atha Yogānuśāsanam.

Now (begins) the instruction on Yoga.

Comments

The uses of the words *yoga* and *anuśāsana* in the very first aphorism of the *Yogasūtra* are not without the background in the works of the Buddhist meditation theorists, particularly Asaṅga and Vasubandhu.

(1) Yoga

The word *yoga* is used in various senses. In the present context, the word *yoga* can be taken to mean either of two things:

(1) The state of self-control and meditative absorption.
(2) The practice of self-discipline which has meditation as its core.

Patañjali uses the word yoga in these senses, and the word was used in one of these senses in the *Upaniṣads*[1] also. But it is incorrect to conclude from this that Patañjali was carrying forward the legacy of the Vedas and *Upaniṣads* in a simple and straightforward way. This is because around the second to fourth centuries, Buddhist thinkers like Maitreya and Asaṅga adopted the term Yoga and made it the central concept of their formulation of Buddhist soteriology. In *Śrāvakabhūmi*, Asaṅga explains words like yoga, *yogācāra* and *yogabhāvanā* with regard to a connotation of the path of meditation.[2]

Patañjali's use of the word *yoga* here can be said to be continuous with this but also a response to it.

(2) Anuśāsana

Instruction. The aphorism "*atha yogānuśāsanam*" is compared by some to the opening statement of *Vyākaraṇamahābhāṣya*, namely, "*atha śabdānuśāsanam*" ("Now we begin the instruction in (the science) of word") to indicate that the two Patañjalis, authors of *Yogasūtra* and *Vyākaraṇamahābhāṣya*, may be the same.[3] However, the similarity of the opening statements is weak (and unconvincing) evidence for the sameness of the two authors. If, on the other hand, the *Yogasūtra*'s opening aphorism is dissociated from the context of *Vyākaraṇamahābhāṣya* and considered in light of its Buddhist background, the word *anuśāsana* can be understood differently. The word has special significance in the context of the Buddha's teaching. Vasubandhu refers to the three ways, or three *prātihārya*s (marvels), the Buddha used to attract the minds of followers: supernormal power (*ṛddhi*), mind reading (*ādeśana*) and instruction (*anuśāsana*). Of these, *anuśāsana* is regarded as the best because it does not deviate from the truth (*avyabhicāritvāt*) and is associated with the desired beneficial fruit by instructing the remedy.[4] Patañjali might have had some such idea while using the word in the first aphorism.

Vyāsa's commentary of YS I.1 is important from the point of view of the Buddhist influence. Vyāsa introduces the doctrine of *cittabhūmi*s (planes of mind) and suggests that *samādhi*, that is, concentration, is common to all. These ideas are rooted in Buddhist Abhidharma literature. Vasubandhu refers to a sixteen-fold classification of *citta* provided by Vaibhāṣikas, which they broadly divide into two groups: defiled (*kliṣṭa*) and wholesome (*kuśala*).[5] The classification includes all five varieties taken up by Vyāsa as five planes (*bhūmi*): *kṣipta* (thrown, disturbed), *līna* (the same as *mūḍha*, influenced by indolence, inertia, *kausīdya*), *vikṣipta* (which focuses outside of the five objects of senses)[6] and *samāhita* (same as *ekāgra*, one-pointed). Vasubandhu's classification is given in tabular form in Table 1.1.

It is clear from the Table that *vikṣipta* and *līna* of the Buddhist Abhidharma match *vikṣipta* and *mūḍha* of Vyāsa's classification, while *samāhita-citta* of the Buddhist Abhidharma matches *ekāgra* of Vyāsa's classification. *Vimukta* is comparable with *niruddha* of Vyāsa's classification.

Similarly, the distinction between *kṣipta* and *vikṣipta* is clearly made by Vasubandhu. He defines *kṣipta* in following words: "The mind which is restless due to elemental imbalance caused by *karma*, and which has lost control and alertness, is called *kṣipta*".[7] In the same context, Vasubandhu also distinguishes between *kṣipta* and *vikṣipta*. Here Vasubandhu introduces four possibilities (*catuṣkoṭi*):

(1) Mind may be *kṣipta*, but not *vikṣipta*. (It may be disturbed and unstable and not even temporarily concentrated.)

TABLE 1.1 Vasubandhu's classification of mind (*citta*) into two groups: defiled (*kliṣṭa*) and wholesome (*kuśala*)

Defiled type of mind	*Wholesome type of mind*
vikṣipta (externally focused)	*saṅkṣipta* (internally temporarily focused)
līna (indolent)	*pragṛhīta* (energetic)
parītta (little, mean)	*mahadgata* (lofty)
uddhata (agitated)	*anuddhata* (balanced, not agitated)
asamāhita (unstable, not concentrated)	*samāhita* (stable, concentrated)
avyupaśānta (disturbed)	*vyupaśānta* (undisturbed, peaceful)
abhāvita (non-contemplative)	*bhāvita* (contemplative)
avimukta (not emancipated)	*vimukta* (emancipated)

Source: AKB VII.11d

(2) Mind may be *vikṣipta*, but not *kṣipta*. (It may be defiled and temporarily concentrated, but not disturbed.)
(3) Mind may be both *kṣipta* and *vikṣipta*. (It may be both defiled and disturbed.)
(4) Mind may be neither *kṣipta* nor *vikṣipta*. (It may be fully concentrated; neither defiled nor disturbed.)

Buddhist classification of *citta*s (whether sixteen-fold or twenty-fold) is not referred to as the planes of the mind, because Buddhists do not arrange them hierarchically. Though a broad hierarchy of defiled and undefiled (or unwholesome and wholesome) minds is indicated by Buddhists, as we have seen, the types of defiled minds themselves are not arranged in this way. Vyāsa arranges defiled minds—namely *kṣipta*, *mūḍha* and *vikṣipta*—in a hierarchy. Similarly, he arranges *ekāgra* and *niruddha* in hierarchical order. Hence, he calls them planes, and not just kinds, of mind.

Another significant point in this context is that according to Vyāsa, *yoga*—that is, *samādhi* (which is the same as *ekāgratā*)—is the common characteristic of mind belonging to all *bhūmi*s ("*sārvabhaumaś cittasya dharmaḥ*"). Here the word *bhūmi* can be taken to mean "field or jurisdiction". This idea is also based on the Buddhist theory of mind. Here one-pointedness of mind is called *mahābhūmika* ("having a great field"). Vasubandhu says:

> Field (*bhūmi*) means the object (or region) of movement (*gativiṣaya*). Whatever is the object (or region) of movement of something is called its field. Those which have a great field are called "*mahābhūmika*" as they are present in all types of mind.
>
> *(AKB II.23)*

Vasubandhu here refers to ten factors as common to all minds: feeling (*vedanā*), volition (*cetanā*), perception (*saṁjñā*), wish (*chanda*), contact (*sparśa*), thought (*mati*), mindfulness (*smṛti*), attention (*manasikāra*), determination (*adhimokṣa*) and concentration (*samādhi*) (AK II.24). Vyāsa is concerned not with all the common mental factors mentioned by Vasubandhu, but rather with only *samādhi* as belonging to all types of mind.

Vyāsa here raises the question of whether *samādhi*, if it belongs to all types of mind, will also belong to the (externally) focused mind ("*vikṣipte cetasi*") as well. His answer is that in the (externally) focused mind, the characteristic of one-pointedness gets subordinated by defilement and hence it does not attain the status of "*yoga*". Vyāsa's approach echoes Vasubandhu, who says: "Since *samādhi* is *mahābhūmika* (that which has great jurisdiction), all minds will have to be regarded as one-pointed. This is not so because (in other states) one-pointedness is weak".[8]

The stage/plane of mind called "ceased" (*niruddha*) can be compared with the Buddhist idea of *nirodha-samāpatti*. The latter is a meditative state in which all mental factors such as *saṁjñā* and *vedanā* have ceased. Vasubandhu discusses whether, in the state of *nirodha-samāpatti*, *citta* also ceases. He points out that for Vaibhāṣikas, *citta* also ceases.[9] The question then is how the aspirant can come out of that stage. How can the mind-series, once stopped for a while, commence again? One answer is: mind can arise again from the body coupled with the senses.[10] The other answer comes from Bhadanta Vasumitra, who opines that there is no cessation of mind in *nirodha-samāpatti*.[11] Vyāsa's conception of *niruddha-citta-bhūmi* is closer to Vasumitra's position, because although in this *bhūmi* all mental states (*citta-vṛttis*) have ceased, the pure mind continues to exist.

योगश्चित्तवृत्तिनिरोधः ।।2।।

Yogaś cittavṛttinirodhaḥ.

Yoga is cessation of the modifications of mind.

Comments

This aphorism gives the definition of *yoga* as *citta-vṛtti-nirodha*. The terms *citta*[12] and *nirodha* are not found in Sāṅkhya literature. However, they are central terms of Buddhist psychology and soteriology. Patañjali seems to have adopted the Buddhist terms and used them for explaining the final state of *samādhi*.

Citta

Though Patañjali adopts the Buddhist term *citta*, he tries to fit it into the Sāṅkhya metaphysical framework, where *puruṣa* is accepted as the abode of consciousness. Hence, the concept of *citta* cannot become supreme in Patañjali's system as it is in Buddhism. Two points of difference between the two conceptions of *citta* become important in this context.

(1) In Buddhism, *citta* is essentially conscious. If it is located in a Sāṅkhya framework, it will belong to the realm of *prakṛti* and hence will be regarded as insentient.
(2) In Buddhism, no one like *puruṣa/ātman* is accepted. In Sāṅkhya, *puruṣa* is accepted as existing independently of *citta*. Though the classical Sāṅkhya does not identify *puruṣa* with *ātman*, it is under constant pressure from Upaniṣadic systems to do so.[13]

Many *Yogasūtra* commentators have tried to reduce Patañjali's Yoga to a kind of Sāṅkhya and hence have interpreted *citta* as either *buddhi* (*mahat*) or *antaḥkaraṇa*. Patañjali himself does not seem to be favourable to such reductionism.

Perhaps by using the word *citta*, Patañjali wants to slightly deviate from Sāṅkhya and suggest that in Yoga, we are not dealing with the insentient *buddhi* of the classical Sāṅkhya, but with something sentient or conscious. Hence *samādhi*, according to Patañjali, is the cessation of modifications of the conscious mind. This position is closer to Buddhism.

In addition to *citta*, Patañjali also accepts *puruṣa*, which is supposed to be a typically Sāṅkhya concept. But unlike Sāṅkhya, Patañjali brings the two notions, namely *citta* and *puruṣa*, closer to each other. Patañjali on many occasions avoids use of the term *puruṣa*, which refers to a metaphysical entity, and uses terms of psychological or phenomenological character instead. For example, he uses the words *draṣṭṛ* (YS I.3, II.17, II.20, IV.23), *dṛṣi* (II.20, II.25), *dṛkśakti* (II.6), *grahītṛ* (I.41), *citi* (IV.22) and *citiśakti* (IV.34) for *puruṣa*. How this brings *puruṣa* and *citta* closer to each other can be seen in Table 1.2.

Puruṣa, of the Sāṅkhya system, is a transcendent metaphysical entity. But when it is re-termed, *puruṣa* becomes more empirical or experiential. While Sāṅkhya is concerned with discrimination between *prakṛti* and *puruṣa* as the final goal to be achieved through speculative reasoning, Patañjali often talks of discrimination and correlation between *dṛkśakti* and *darśanaśakti* (YS II.6), or between *sattva* and *puruṣa* (YS III.35, 49, 55), both of which are "psychic" in nature. The point is that even when Patañjali accepts the Sāṅkhya metaphysical framework, he moulds it into a psychological form which can share many things with Buddhism.

Citta-vṛtti

The term *citta-vṛtti*, which means "mind's becoming (something)" or "modification of mind" or "mental state", does not occur in the Buddhist literature. However, another term, *caitta* (or *cetasika*, literal meaning: "mental", generally translated as mental factor), occurs which has almost the same meaning. Abhidharma Buddhism talks about different minds and mental states (*cittas* and *caittas*); Vasubandhu observes that in the highest state of meditative trance (which is called *saṁjñā-vedayita-nirodha*: cessation of perception as well as sensation), there is *citta-caitta-nirodha* (cessation of minds and mental states).[14] Patañjali, however, is talking here of *citta-vṛtti-nirodha* (which is parallel to *caittanirodha* of Buddhist Abhidharma), and not *cittanirodha*. The difference is significant. For Vaibhāṣika and Sautrāntika Buddhists, there is no *citta* without *caitta* and no *caitta* without *citta*.[15] So, if one of them ceases, the other also should cease. For Patañjali, on the other hand, *citta-vṛtti* is a modification of *citta*, but there can be pure *citta* without modification. In meditative trance, there is pure *citta* without modification.

There is another angle to this issue. In its literal sense, the word *vṛtti* means "being". Hence, *citta-vṛtti* means "being of a *citta*". In that case, the classification

TABLE 1.2 Phenomenological translation of Sāṅkhya terminology

Words used for puruṣa	*Words used for* citta
draṣṭṛ, dṛṣi, dṛkśakti	*Darśanaśakti*
grahītṛ	*grahaṇa*
citi, citiśakti	*Citta*

Sources: YS I. 2, 3, 41; II.6, 17, 20; IV. 16, 17, 22, 23, 34, etc.

of *citta* will amount to the classification of the different ways in which *citta* is. Thus, the Buddhist classification of *citta* overlaps with that of *citta-vṛtti*s.

Nirodha

In classical Sanskrit, *nirodha* is generally taken to mean suppression. If we accept this meaning in the context of Patañjali's Yoga, then the definition of "*yoga*" as "*citta-vṛtti-nirodha*" will be understood as "suppression of the states of mind". Against this, *nirodha* in the Buddhist context means cessation. This meaning throws better light on the Pātañjala-yoga definition of *yoga*. *Yoga* as the state of meditative absorption makes better sense if it is understood as the cessation (which can be natural or spontaneous) of mental states rather than their suppression or restriction (which is generally deliberate or forced).[16]

I want to suggest that the classical Sanskrit word *nirodha* should be distinguished from the Buddhist Sanskrit (that is, originally Pali) word *nirodha*. The classical Sanskrit word is derived from the transitive verb *ni-rudh*, which means to hold back, stop, hinder, shut up, confine, restrain, check, suppress and destroy.[17] The Buddhist Sanskrit (that is, originally Pali) word *nirodha* is derived from the Pali verb *nirujjhati*, which operates as an intransitive or passive verb and means to be broken up, to be dissolved, to be destroyed, to cease, to die.[18] *Nirodha* in Buddhism as "stoppage" is also intransitive ("y stops"), not transitive (of the form "X stops y"). Patañjali's use of the word *nirodha* should be understood as on par with the Buddhist use.

तदा द्रष्टुः स्वरूपेऽवस्थानम् ।।3।।

Tadā draṣṭuḥ svarūpe'vasthānam.

At that time the seer (the spirit-*puruṣa*) abides in his own nature.

Comments

Cessation of suffering can be called the ultimate goal. In Sāṅkhya, it is referred to as *kaivalya* and in Buddhism it is called *nirvāṇa*. *Kaivalya* in Sāṅkhya is generally understood as a transcendental metaphysical state, whereas *nirvāṇa* in Buddhism is a more empirical–psychological state, though Buddhists also try to give it a transcendental status. Patañjali avoids the word *nirvāṇa*, but he talks about the ultimate goal at both the empirical and transcendental levels.

It is interesting to consider more carefully the description of the ideal state which appears here. The ideal state has two aspects: (1) "Yoga is the cessation of mental states" (YS I.2); (2) "In that state the seer (witness consciousness) remains as he is" (YS I.3).

The first aspect can be called psychological in nature, whereas the second aspect is of transcendental–metaphysical. The second aspect, which refers to *draṣṭā*, that is, *puruṣa*, is dominated by Sāṅkhya. However, the first aspect, which contains *citta* and *nirodha* as the central concepts, is definitely dominated by Buddhism.

In the commentary, Vyāsa uses the term *vyutthāna-citta*, which refers to the state of mind when the mind has come out of *samādhi*. The term seems to be based on the Buddhist theory of meditation.[19]

वृत्तिसारूप्यमितरत्र ।।4।।

Vṛttisārūpyam itaratra.

On other occasions the Seer becomes similar in form as the mental state.

Comments

Although the Seer in himself does not undergo any modification, it appears to undergo modifications in the states other than *samādhi* (that is, in the *vyutthāna* state). This is a typically Sāṅkhya explanation.

Vyāsa explains the ambivalent state of the *puruṣa* in the *vyutthāna* state by using the metaphor of a magnet (*ayaskāntamaṇi*) for the mind. The seer is in fact isolated and does not own anything. But it is (as if) the mind, through its magnetic power, attracts the seer and becomes his property; it is (as if) the seer becomes the owner of the mind, so that what belongs to the mind is also taken to belong to the seer, and the seer is taken to assume the form of the mind as its own form. Hence, if the mind is happy, the seer is taken to be happy and so on. Vyāsa is explicitly following the line of Sāṅkhya.

वृत्तयः पञ्चतय्यः क्लिष्टाक्लिष्टाः ।।5।।

Vrittayaḥ pañcatayyaḥ kliṣṭākliṣṭāḥ.

Mental states are of five kinds (which are) either defiled or non-defiled.

Comments

Here Patañjali clubs together two different classifications of mental states. The five-fold classification, which will be given in the next aphorism, is epistemological, whereas the two-fold classification introduced here is moral–psychological or soteriological. The two-fold classification given here reflects the similar Buddhist classification of *citta*, which is also referred to as that between *sāsrava* and *anāsrava*. (*Āsrava* means an evil influence caused by past action. Similarly, *kleśa* means defilement or impurity caused by past action. *Sāsrava-citta* or *kliṣṭa-citta* therefore means the mind containing such an evil influence or impurity. *Anāsrava-citta* or *akliṣṭa-citta* means the mind which lacks such an impurity.)

In his commentary, Vyāsa explains *kliṣṭa* (defiled) mental states as those caused by *kleśa*s and which are the loci (*kṣetra*, literally, fields) of the collection of *karmāśaya* (*karmāśayapracayakṣetrībhūtāḥ*). The notion of *karmāśaya* is similar to the notion of *karmāsrava*, which means evil influence of an action or an impurity caused by an action. The word *āsrava* occurs in both Jaina and Buddhist traditions with the same import as karmic influence with slightly different connotations.

In Jainism, *āsrava* means the inflow of karmic particles into *jīva*. In Buddhism, it means influence of action on mind. The difference in connotation is mainly due to the difference in the metaphysical understanding of karma and mind/self. The main point is that defiled mental states lead to a continuation of the birth cycle whereas non-defiled mental states lead to emancipation. Vyāsa is following this line, except he is explaining the state of bondage or the birth cycle in terms of the operation of the strands of *prakṛti* (*guṇādhikāra*) and emancipation in terms of *khyāti* (discriminative knowledge).

प्रमाणविपर्ययविकल्पनिद्रास्मृतयः ।।6।।

Pramāṇaviparyayavikalpanidrāsmṛtayaḥ.

(The five mental states are:) true cognition, error, verbal cognition, (deep) sleep and recollection.

Comments

This is a typically epistemological classification of "mental states". Though it is partly consistent with Sāṅkhya epistemology, it does not seem to be derived from the latter. It is partly consistent with the Vaiśeṣika classification of *vidyā* and *avidyā*. It is partly consistent with the Upaniṣadic classification (as in *Māṇḍūkyopaniṣad*) of the states of *ātman* and also consistent with Buddhist epistemology. But it is not fully consistent with or directly derived from any one of them.

Sāṅkhya

Similarity with Sāṅkhya: Three-fold classification of *pramāṇa*: perception, inference and verbal testimony.

Dissimilarity: There is no explanation of erroneous cognition (*viparyaya*) in general in Sāṅkhya. Though the term *viparyaya* is used in *Sāṅkhyakārikā*, it refers to metaphysical illusion or misconception rather than illusion in general. Similarly, the concept of *vikalpa* (mental construction) is absent in Sāṅkhya.

Vaiśeṣika

Similarity with Vaiśeṣika: *Viparyaya* as false cognition is included under *avidyā*. *Smṛti* (recollection) is included under *vidyā* (right cognition). But it is distinguished from *pramāṇa*. *Ārṣa-jñāna* (the knowledge of sages) accepted as *vidyā* in Vaiśeṣika is similar to *āgamapramāṇa* of Yoga.

Dissimilarity: There is no awareness in deep sleep according to Vaiśeṣika. According to Yoga, there is awareness in deep sleep. *Saṁśaya* and *anadhyavasāya* (doubt and indeterminate cognition), accepted as erroneous conditions in Vaiśeṣika, are not so recognized in Yoga. Similarly, *vikalpa* has no role to play in Vaiśeṣika epistemology.

Buddhism

The Buddhist classification of mental states is more of a moral–psychological nature than epistemological. Hence, the concept of *vikalpa* (mental construction) has a significant role to play in Buddhism. Of course, Buddhism does have epistemological elements which are scattered. Those elements will be pointed out while considering the specific kind of *citta-vṛtti*.

Māṇḍūkyopaniṣad

Similarity with *Māṇḍūkyopaniṣad* : In *Māṇḍūkyopaniṣad*, deep sleep (*suṣupti*) is regarded as a state of *ātman*. Accordingly, a level of awareness exists even in that state. This is similar to the Yoga view that deep sleep is a mental state.

Dissimilarity: Yoga is talking about states of mind rather than *ātman*.

प्रत्यक्षानुमानागमाः प्रमाणानि ।।7।।

Pratyakṣānumānāgamāḥ Pramāṇāni.

True cognitions are: perception, inference and (knowledge by) verbal testimony.

Comments

The Sāṅkhya system is known for its epistemology of three *pramāṇa*s, namely perception, inference and verbal testimony. Patañjali can be said to be following Sāṅkhya here. But his position is also not far removed from Buddhism.

From Diṅnāga onwards, the Buddhist epistemologists accepted the model of "two and only two *pramāṇa*s". But the Buddhist epistemology before Diṅnāga seems to broadly accept the model of three sources of knowledge: perception, inference and verbal testimony. This view seems to be shared by *Upaniṣads* and Sāṅkhya as well as Buddhism in the early period.[20] Vasubandhu often uses the dual method of *yukti* (reason) and *āgama* (verbal testimony) for justifying his views.

Vyāsa's explanation of the three *pramāṇa*s seems to be influenced by Buddhist epistemology. The idea that the object has two aspects—universal and particular (*sāmānyalakṣaṇa* and *svalakṣaṇa*)—is found in Vasubandhu. Vyāsa holds that out of the two characteristics, namely specific (*viśeṣa*) and universal (*sāmānya*), perceptual cognition grasps mainly the specific aspect (*viśeṣāvadhāraṇapradhānā vṛttiḥ pratyakṣam*) and inferential cognition grasps mainly the universal aspect (*sāmānyāvadhāraṇapradhānā vṛttir anumānam*). This idea is close to the epistemology of Diṅnāga and Dharmakīrti, according to which the object of perception is a particular whereas the object of inference is a universal. The roots of this epistemological approach in Vasubandhu or other earlier Buddhist epistemologists need to be traced.

विपर्ययो मिथ्याज्ञानमतद्रूपप्रतिष्ठम् ।।8।।

Viparyayo mithyājñānam atadrūpapratiṣṭham.

Error is the false cognition, based on what is not the nature of the object.

Comments

It is difficult to trace the source of this definition, but the notion of *viparyaya* is commonly used in later epistemological works. See, for instance, the definitions of *viparyaya* in *Praśastapādabhāṣya*: "Error is cognising something as that, what is not that"; "Error is a false cognition".[21]

Vyāsa, in his commentary, points out that the erroneous cognition is sublated by true cognition. He gives the example of "two moon perception" (a case of illusion), which is falsified by the veridical perception of the single moon. This is well taken. But immediately after that, he identifies *viparyaya* with the five-fold *avidyā*, which he identifies with *kleśa*s of YS II.3. He subsequently identifies it with the five-fold *viparyaya* of *Sāṅkhyakārikā*. All this is both uncalled for and misleading.

Here it is necessary to note that Patañjali's two classifications of *citta-vṛttis* belong to different categories. The two-fold classification of defiled and non-defiled belongs to the category of moral–spiritual psychology or soteriology. The five-fold classification belongs to epistemology. Possessing *avidyā* in a soteriological sense and having a perceptual illusion in an epistemological sense are two different things. Vyāsa is mixing up the two categories here.

शब्दज्ञानानुपाती वस्तुशून्यो विकल्पः ।।9।।

Śabdajñānānupātī vastuśūnyo vikalpaḥ.

Conceptual construction is that which follows cognition of a word but which is devoid of a real object.

Comments

Patañjali accepts verbal cognition as cognition different from both "true cognition" (*pramāṇa*) and "illusory cognition" (*viparyaya*). When words are uttered in the course of communication, certain ideas arise in the mind of the listener. These ideas are not real things, and hence the cognition that the listener has is not a true cognition in a strict sense of the term. But the communication generated by it can create authentic practices. Thus, the cognition arising from words cannot be outright rejected as false.

Vikalpa (mental construction or conceptual construction) is an important concept in Buddhism. Though in Buddhism we find degrees of realism and idealism, an idea common to different forms of Buddhism is that, under the spell of ignorance, we construct some objects mentally; and though these constructions serve practical purpose, the liberated state also involves liberation from mental constructions.

Asaṅga in *Bodhisattvabhūmi* (Chapter IV: *Tattvārthapaṭala*) underlines this nature and role of *vikalpas*: "Ignorant people do not understand the true nature of things. As a result of that arise eight types of mental constructions (*vikalpa*) which create three kinds of things".[22] The eight mental constructions (or ideas) are (examples are given in brackets): (1) natural idea ("material object"); (2) specific idea ("visible material object", "wholesome object"); (3) the idea of collection ("forest", "cloth"); (4) the idea of "I" ("I"); (5) the idea of "mine" ("this is mine"); (6) the idea of "liked" ("this is good/beautiful"); (7) the idea of "disliked" ("this is bad/ugly"); and (8) the idea of neither liked nor disliked ("this is neither good nor bad"). These eight types of mental constructions give rise to three types of things as follows: (1) material objects (*rūpa* etc.); (2) the notion of self; and (3) unwholesome roots (desire-aversion-delusion). Asaṅga again says that mental constructions cause their objects (*ālambana*), and objects in their turn cause mental constructions; this process has been going on without beginning.

It is doubtful whether Patañjali has all this in mind when he accepts *vikalpa* as a mental state, different to both *pramāṇa* and *viparyaya*. His position becomes ambivalent because, in his metaphysical views, he is closer to the realist Sāṅkhya, whereas in his psychology and epistemology of meditation, he is close to Buddhism.

In his examples of *vikalpa*, Vyāsa tries to protect Sāṅkhya realism.[23] He restricts the scope of *vikalpa* considerably. It is doubtful whether Patañjali wished to restrict it so much.

It is interesting to note here that Vasubandhu (AKB I.33) identifies "natural construction" with thought (*vitarka*). Hence *nirvitarka-samādhi* gets identified with *nirvikalpa-samādhi*.

अभावप्रत्ययालम्बना वृत्तिर्निद्रा ।।10।।

Abhāvapratyayālambanā vṛttir nidrā.

Deep Sleep is the mental state, the object of which is the cognition of absence.

Comments

Except Nyāya-Vaiśeṣika and Cārvāka, all schools of Indian philosophy maintain that there is some degree of consciousness even in the state of deep sleep. Sāṅkhyas would say that there is a manifestation of *buddhi* in that state, though it is dominated by *tamas*. Buddhists would say that the mind (which marks awareness) does exist in that state, though it is dominated by the hindrance called *middha* (sleepiness, drowsiness). Buddhists regard the mind continuum as continuous during deep sleep. Vasubandhu (AKB VIII.16) says that "The mind-continuum continues according the intention of persons. For example, one who makes a resolve before sleeping that he will get up at such and such a time, gets up accordingly".

अनुभूतविषयासम्प्रमोषः स्मृतिः ।।11।।

Anubhūtaviṣayāsampramoṣaḥ smṛtiḥ.

Recollection means "non-removal of the experienced object".

Comments

Patañjali in this aphorism defines *smṛti* as a distinct kind of mental state. *Smṛti* in the epistemological context means recollection or memory cognition. Most epistemological systems in India accept recollection as a kind of cognition but do not include it under *pramāṇa*. The main reason is that recollection does not give any new information. It repeats what is already cognized. The same policy might have been adopted by Patañjali while accepting *smṛti* as a *citta-vṛtti* other than *pramāṇa*. But what is the nature of this *smṛti*? Here Patañjali is under the Buddhist influence.

Smṛti in the sense of memory cognition is supposed to be the result of the awakening of an impression (*saṁskāra*) generated by an experience.[24] Patañjali does not define *smṛti* in that way. The term *smṛti* has a different connotation in Buddhist psychology. It is that of attentiveness or mindfulness of mind. (Though the two connotations of the term, memory and mindfulness, are not totally unrelated, they are different.[25]) Vasubandhu gives the definition of *smṛti* as "*ālambanāsampramoṣaḥ*"[26] (non-removal of the object from mind). Patañjali's definition of *smṛti* as "*anubhūtaviṣayāsampramoṣaḥ smṛtiḥ*" ("non-removal of the experienced object") follows the Buddhist definition.

It is interesting to note that *sampramoṣa* seems to be a Hybrid Sanskrit word, derived from the Pali word *sampamosa*. The Pali word is derived from the root *sampamussati*, the Sanskrit analogue of which is *sam+pra+mṛṣ* (meaning to wipe, clean). In Buddhist Sanskrit, the transformed word *sampramoṣa* got reinterpreted as one derived from the root *sam+pra+muṣ*, which means to steal.

When the word entered the *Yogasūtra*, the commentators, forgetting its association with Hybrid Sanskrit or Pali words, interpreted it as derived only from the root *sam+pra+muṣ*.

अभ्यासवैराग्याभ्यां तन्निरोधः ।।12।।

Abhyāsavairāgyābhyāṃ tannirodhaḥ.

The cessation of them (i.e. of the modifications of mind) is brought about by repeated practice and detachment.

Comments

Patañjali says that the Yoga (cessation of the mental states) can be achieved through *abhyāsa* and *vairāgya*. These ideas are vividly seen in the Buddhist literature, though in a scattered manner.

In Buddhist soteriology, two paths are acknowledged, *darśanamārga* and *bhāvanāmārga*. *Darśanamārga* is the path of perception or realization. *Bhāvanāmārga* is the path of cultivation of mind through meditative practice. In this context, *abhyāsa* (continuous/repeated practice) is closely connected with *bhāvanāmārga*. In fact, Asaṅga defines *bhāvanā* as a kind of *abhyāsa*.[27]

Vairāgya (detachment) is the absence of *rāga* (attachment). Detachment is attained and developed by pursuing both paths, that is, the path of realization and the path of cultivation (meditation).

(1) Through the path of realization: when one realizes that the empirical phenomena are unsatisfactory (*duḥkha-darśana*), then naturally one abandons attachment towards them.

In this sense, *rāga* is abandonable through the realization of un-satisfactoriness (*duḥkha-darśanaheya*).

(2) Through the path of cultivation: through the realization of the unsatisfactory (i.e. *duḥkha*) character of things and other noble truths (such as the cause of suffering, the cessation of suffering and the path leading to cessation), one may abandon attachment. But in order to develop this detachment, one has to practice the path repeatedly. As Vasubandhu says:

> Four (defilements) namely attachment, hatred, conceit and misconception are abandonable through meditation (*bhāvanāheya*) because they are abandoned by repeated practice of the path (*mārgābhyāsa*).[28]

In this way, *vairāgya* is partly a product of *abhyāsa*. But when it is achieved, it becomes almost a separate means to liberation.

Spiritual development is caused by *adhobhūmi-vairāgya* (detachment towards the objects belonging to the lower realm). When one sees faults (*ādīnava*) in the objects which belong to the present realm, one develops detachment towards them and goes to the higher realm; ultimately, one develops detachment towards objects belonging to all the realms. This will be clear in the comments on YS I.15 and I.16.

तत्र स्थितौ यत्नोऽभ्यासः ।।13।।

Tatra sthitau yatnobhyāsaḥ.

Repeated practice is the effort for remaining steady in it.

Comments

The commonly accepted meaning of the term *abhyāsa* is repeated practice. In Buddhist soteriology, the word is used mainly in two contexts: efforts for accumulating religious merits (*puṇya*) and the practice of meditation (*bhāvanābhyāsa*). Patañjali's use of the term *abhyāsa* seems to have specific reference to meditation.

स तु दीर्घकालनैरन्तर्यसत्कारासेवितो दृढभूमिः ।।14।।

Sa tu dīrghakālanairantaryasatkārāsevito dṛḍhabhūmiḥ.

It (i.e. practice) becomes firmly grounded on being continued for a long time, without interruption and with devotion.

Comments

Patañjali here talks about the way in which *abhyāsa* has to be done: according to him, *abhyāsa* becomes firmly rooted when it is practiced for a long time without gap and with felicitation (or devotion). In *Abhidharmakośabhāṣya*, Vasubandhu refers to these characteristics in relation to Buddha's *abhyāsa* of accumulation of the wealth of merit and knowledge.[29]

So, these ideas must have been in the Buddhist tradition, which Patañjali crystallizes in an aphorism.

दृष्टानुश्रविकविषयवितृष्णस्य वशीकारसंज्ञा वैराग्यम् ।।15।।

Dṛṣṭānuśravikaviṣayavitṛṣṇsya vaśīkārasaṃjñā vairāgyam.

Detachment is the consciousness of mastery (over desire) in one who is free from craving for empirical (i.e. this-worldly) as well as scriptural (i.e. other-worldly) objects.

Comments

Patañjali is here introducing the preliminary concept or ordinary concept of detachment. He defines it as the consciousness of mastery (or the conscious act of overpowering) over desires or over the objects of desire. Such mastery arises by becoming free from cravings (*vitṛṣṇa*) towards empirical and scriptural objects.

In Buddhist soteriology, the notion of mastery (over objects of desire) occurs in terms of *abhibhvāyatana* (the sphere of mastery). The term *abhibhū*, derived from the root *abhi+bhū*, means "to overcome", "to master". Vasubandhu refers to the technique of observing objects one by one with the sense of overcoming them[30] (that is, not getting attached to them.) It is possible that Patañjali's notion of *vaśīkārasaṁjñā* has something to do with the idea of *abhibhvāyatana*.

तत्परं पुरुषख्यातेर्गुणवैतृष्ण्यम् ।।16।।

Tat paraṃ puruṣakhyāter guṇavaitṛṣṇyam.

The higher form of detachment is freedom from cravings for the strands (of *prakṛti*) which is attainted through the (discriminating) knowledge of the Self (*puruṣa*).

Comments

Patañjali classifies detachment (*vairāgya*) into lower and higher (*apara* and *para*). It seems to be a reformulation of the Buddhist ideas of "*laukikena mārgeṇa vairāgyam*" and "*lokottareṇa mārgeṇa vairāgyam*" (detachment by mundane way and detachment by supra-mundane way).

The general idea behind the Buddhist classification is that when the practitioner wants to attain a meditative trance belonging to a higher plane (*bhūmi*), he has to develop detachment towards the objects belonging to lower planes.[31] When he reaches the top, he has developed detachment towards all objects belonging to all planes.

Patañjali presents this idea in abridged form in the aphorisms (YS I.15–16). However, he presents the highest *vairāgya* in Sāṅkhya terminology ("*guṇavaitṛṣṇyam*").

According to Sāṅkhya, all objects are manifestations of the three strands of *prakṛti*. Hence, Patañjali seems to maintain that someone who develops detachment towards the three strands automatically develops detachment towards all objects.

वितर्कविचारानन्दास्मितारूपानुगमात्संप्रज्ञातः ।।17।।

Vitarkavicārānandāsmitārūpānugamāt saṃprajñātaḥ.

The object-conscious meditative absorption arises by means of accepting forms of gross objects of thought, subtle objects of thought, joy and I-notion (in their sequential order).

Comments

Patañjali classifies *samādhi* into two kinds: object-conscious (*samprajñāta*) and the other (*anya*). Here he is defining the first type (*samprajñāta*) as a kind of meditative absorption which follows certain forms (*rūpānugama*). Here he mentions four forms: gross thought-objects, subtle thought-objects, joy and I-notion. Patañjali's use of the word *rūpa* in the *sūtra* and his mention of *vitarka*, *vicāra* and *ānanda* clearly indicate the link between his notion of Buddhism's *samprajñāta*[32] *samādhi* and *rūpadhyāna*. In *rūpadhyāna* accepted in the Buddhist theory of meditation, the following forms are primarily included: *vitarka-vicāra-prīti-sukha-ekāgratā*. In Patañjali's scheme, *prīti* and *sukha* seem to be clubbed into one (*ānanda*) and *ekāgratā* seems to be replaced by *asmitā*.

In the Buddhist scheme, the earlier forms drop out when the form-meditation progresses.[33] In his commentary, Vyāsa maintains this idea of dropping of earlier forms one by one, with some difference in details. The Buddhist scheme is presented in terms of dropping of factors in Table 1.3.

In Vyāsa's scheme, there is also dropping of factors in the same order (but with the elements slightly altered; see Table 1.4).

Here two differences should be considered:

(1) In the Buddhist scheme, it appears that *vitarka* and *vicāra* both drop out together, whereas in Vyāsa's scheme first *vitarka* drops out and then *vicāra*. This difference is not significant, because in the Buddhist scheme as well,

TABLE 1.3 The stages in *rūpadhyāna*

	Elements	*Dropping of -*
1st *dhyāna*	*vitarka-vicāra-prīti-sukha-ekāgratā*	-
2nd *dhyāna*	*prīti-sukha-ekāgratā*	*vitarka-vicāra*
3rd *dhyāna*	*sukha-ekāgratā*	*Prīti*
4th *dhyāna*	*Ekagrata*	*Sukha*

Source: AK and AKB VIII.7–8

TABLE 1.4 The stages of *samprajñāta-samādhi*

	Elements	*Dropping of -*
1st *samādhi*	*vitarka-vicāra-ānanda-asmitā*	-
2nd *samādhi*	*vicāra-ānanda-asmitā*	*Vitarka*
3rd *samādhi*	*ānanda-asmitā*	*Vicāra*
4th *samādhi*	*Asmitā*	*Ānanda*

Source: YB I.17

Vasubandhu refers to an intermediate state which he calls intermediate *dhyāna* (*dhyānāntara*).[34] Hence, the intermediate *dhyāna* between Vasubandhu's first and second *dhyāna* of will match with Vyāsa's second *samādhi*.

(2) The other difference is regarding the object of Vasubandhu's fourth *dhyāna* and of Vyāsa's fourth *samādhi*. In Vasubandhu's fourth *dhyāna*, there is *ekāgratā* (one-pointedness) without *prīti-sukha* and so on. In Vyāsa's fourth *samādhi*, there is *asmitā* without *ānanda* and so on. In fact, *asmitā* as the object of a higher stage of *samprajñāta-samādhi* is problematic. In Patañjali's scheme, *asmitā* is a kind of defilement (YS II.3), which is an (erroneous) admixture of *puruṣa* and *buddhi* (*dṛkśakti* and *darśanaśakti*).[35] How does it become a form (or object) of the highest stage of *samprajñāta-samādhi*, then? Vyāsa seems to have realized this problem about *asmitā*. So, in order to protect the Sūtrakāra, he defines *asmitā* in a different way. He says "*ekātmikā saṁvid asmitā*" ("*Asmitā* means uniform consciousness"). Vyāsa here bypasses Patañjali's definition of *asmitā* as a kind of *kleśa* and brings the notion of *asmitā* closer to the idea of *ekāgratā* (one-pointedness).

Vyāsa's definitions of *vitarka* and *vicāra* are also influenced by Vasubandhu's definitions (see Table 1.5).

विरामप्रत्ययाभ्यासपूर्वः संस्कारशेषोऽन्यः ।।18।।

Virāmapratyayābhyāsapūrvaḥ Saṃskāraśeṣo'nyaḥ.

Preceded by the practice of experiencing cessation (of the form-objects) is the other, in which only the latent impressions remain.

Comments

Virāmapratyaya means "experience of stoppage". And *abhyāsa* is meditation practice. Hence, the aphorism means that the other (*anya*) kind of *samādhi* is preceded by concentration on stoppage of the form-objects, and only latent impressions remain in it. Patañjali probably means that in the *samprajñāta-samādhi* one experiences forms (*rūpa*) such as gross thoughts, subtle thoughts, joy and I-notion. In the other (*anya*) type of *samādhi*, one experiences such forms ceasing (*virāma*). This definition can suit *ārūpyadhyāna*, in which there are only formless or empty

TABLE 1.5 Vasubandhu and Vyāsa on *vitarka* and *vicāra*

	Vasubandhu's definition	*Vyāsa's definition*
vitarka	*cittaudārikatā vitarkaḥ* (grossness of mind is *vitarka*)	*vitarkaḥ cittasya ālambane sthūla ābhogaḥ* (*vitarka* is gross application of mind to the object)
vicāra	*citta-sūkṣmatā vicāraḥ* (subtleness of mind is *vicāra*)	*sūkṣmo vicāraḥ* (*vicāra* is subtle application of mind to the object)

Sources: AKB II.33, YB I.17

Note: *Ābhoga* is the term used in Buddhist Sanskrit in the sense of mental effort or application (vide Edgerton's dictionary).

objects such as infinite space (*ākāśānantyāyatana*), (pure objectless) infinite consciousness (*vijñānānantyāyatana*), nothingness (*ākiñcanyāyatana*) and neither perception nor non-perception (*naivasaṁjñānāsaṁjñāyatana*).[36]

Vyāsa calls it *asamprajñāta*; but such a word is not used by Patañjali. In view of the previous explanation, a better nomenclature of this type of *samādhi* would be *arūpasamādhi* rather than *asamprajñāta-samādhi*.

Moreover, Vyāsa identifies this so-called *asamprajñāta-samādhi* with *nirbīja-samādhi* (seedless-absorption), which is again incorrect, because the seed of *saṁskāras* still remains in this *samādhi*. That is why it is called *saṁskāraśeṣa*. It is clear from I.50 and I.51 that in *nirbīja-samādhi* there is cessation of *saṁskāras* also.

भवप्रत्ययो विदेहप्रकृतिलयानाम् ।।19।।

Bhavapratyayo videhaprakṛtilayānām.

It (i.e. the meditative trance) is inborn for the disembodied beings (=gods) and those merged in their original nature.

Comments

Bhavapratya means conditioned by birth. *Videha* means god/deity, and *prakṛtilaya* means the persons/beings who attain the state of merger in *prakṛti* (primordial state). *Prakṛtilaya* is a Sāṅkhya term. It is a state achieved through *vairāgya* according to *Sāṅkhyakārikā*.[37] Such persons have inborn capacity for *samādhi* as explained by Vyāsa.

The relation of gods with *samādhi* is clearer in the Buddhist tradition. One who has achieved *rūpadhyāna* is reborn in a high world called *rūpadhātu*. Similarly, one who has achieved *arūpadhyāna* is reborn in a higher world (*arūpadhātu*) as a god with an inborn capacity for *ārūpyadhyāna*. As Vasubandhu says:

> In the world of form (*rūpadhātu*) the gods in nine stages of three types of meditation remain in that world for a long time with the bliss born from seclusion and concentration—bliss accompanied by joy or bliss without joy. There stages are called "stages of blissful birth" (*sukhopapatti*).[38]

Asaṅga in *Yogācārabhūmi* (YAB: 94–95) gives an account of different gods experiencing the bliss according to the meditative stage in which they are born (see Table 1.6).

श्रद्धावीर्यस्मृतिसमाधिप्रज्ञापूर्वक इतरेषाम् ।।20।।

Śraddhāvīryasmṛtisamādhiprajñāpūrvaka itareṣām.

For others it is preceded by faith, energy, mindfulness, concentration and insight.

Comments

Here Patañjali is referring to the meditative trance which can be achieved by applying means. Patañjali refers to five-fold means: *śraddhā* (faith/confidence), *vīrya* (energy), *smṛti* (mindfulness), *samādhi* (concentration) and *prajñā* (wisdom/insight). In the Buddhist theory of meditation, the same terms are used in the same order but are referred to as the five faculties (*indriya*). The same five are also called five powers (*bala*).

The specific order among the five faculties is important. As Vyāsa explains:

> Faith (*śraddhā*) is the confidence of mind. . . . When the aspirant, desirous of discriminating knowledge, has faith, energy (*vīrya*) is generated in him. In whom energy has arisen, mindfulness (*smṛti*) becomes present. On presence of mindfulness, the mind, which is undisturbed, attains concentration (*samādhīyate*). In the person with concentrated mind arises discriminating wisdom (*prajñāviveka*).[39]

This is a reflection of what Asaṅga says in *Abhidharmasamuccaya*. To state it briefly:

> What is cultivation? By the faculty of faith (*śraddhā*), confidence is created regarding the noble truths. When confidence is created, efforts are made

TABLE 1.6 Blissful births caused by the stages of meditation

Category of gods	*Meditative stage (bhūmi)*	*The kind of bliss they experience*
Gods belonging to form sphere (*rūpāvacara*)	1st *dhyāna* 2nd *dhyāna* 3rd *dhyāna* 4th *dhyāna*	—Joyful bliss arising from discrimination —Joyful bliss arising from concentration —Bliss without joy —Peaceful bliss with pure equanimity and mindfulness
Gods born in formless sphere (*arūpāvacara*)	Formless meditation	Peaceful liberating bliss

Source: YAB: 94–95

for enlightenment through the faculty of energy (vīrya). When energetic efforts are started, mindful awareness is created through the faculty of mindfulness (*smṛti*). When one has mindfulness, one-pointedness of mind is cultivated by the faculty of concentration (*samādhi*). With concentrated mind, one makes the analytical discovery (*pravicaya*) through the faculty of insight (*prajñā*).[40]

तीव्रसंवेगानामासन्नः ।।21।।

Tīvrasaṃvegānām āsannaḥ.

It is proximate for those who are having intense fervour (for concentration).

Comments

This idea has background in Buddhist meditative theory. Vasubandhu, for instance, says "One who produces the stages of fundamental meditation (*mauladhyāna*) with penetrative aspect (*nirvedha*) invariably sees the noble truth in this very life, because he has intense fervor".[41] Vasubandhu uses the word *tīvrasaṁvegatvāt*, which is worth noting.

In his commentary, Vyāsa introduces a nine-fold classification of yogins in terms of *mṛdu*, *adhya* and *adhimātra*. This nine-fold classification reflects the nine-fold classification made by Asaṅga in *Abhidharmasamuccaya* (p. 70) and that by Vasubandhu in AK (VI.33) and AKB on it. Since this refers to Patañjali's classification of means in the next aphorism, it will be discussed in detail under the next aphorism.

मृदुमध्याधिमात्रत्वात्ततोऽपि विशेषः ।।22।।

Mṛdumadhyādhimātratvāt tato'pi viśeṣaḥ.

It is further differentiated on account of being mild, middling and intense.

Comments

Here Patañjali is saying that though a meditative trance becomes proximate if the aspirant has intense fervour, it may not be equally proximate to all. It will be different depending upon the intensity of means. Patañjali gives a three-fold classification of the intensity: *mṛdu* (mild), *madhya* (middling) and *addhimātra* (extreme). The basic idea is very much Buddhistic. Vasubandhu in AK VI discusses different categories of spiritually advanced persons. There are four basic categories (stream-entrant, once-returner, non-returner and *arhat*), but there are many subcategories. While discussing the category of "non-returner" (*anāgāmī*),[42] Vasubandhu says that all non-returners are not on the same level. "The difference among the non-returners is due to the difference in karmas, defilements (*kleśas*) and faculties (*indriyas*)" (AK VII.39). He explains that the non-returner may have accumulated karma to be experienced in this life, next life or some other life. He may have defilements—middle, middling or extensive. Similarly, the faculties of the non-returner could be mild, middling or extensive.

There is striking verbal similarity between the two statements: Vasubandhu says: "*tad**viśeṣaḥ** punaḥ karmakleśendriyaviśeṣataḥ*" (AK VI.39) ". . . ***mṛdvindriyatvāc**ca*".

Patañjali says: "***mṛdumadhyādhimātratvāt** tato'pi **viśeṣaḥ***".

Hence, what Patañjali says seems to be a simplified and generalized version of Vasubandhu.

Vasubandhu classifies *indriya* (faculty) into mild, middling and extensive. Vyāsa replaces *indriya* with *upāya* (means). He expands the three-fold classification into nine-fold. According to him, just as means can be three-fold, the fervour of the yogin can be three-fold. By combining three-fold means and three-fold fervour, we get nine types of yogin. The idea of expanding a three-fold classification into a nine-fold one is found in both Asaṅga and Vasubandhu.[43]

ईश्वरप्रणिधानाद्वा ।।23।।

Īśvarapraṇidhānād vā.

Or, (meditative trance becomes proximate) by making a resolve regarding *īśvara* (the supreme being).

Comments

Īśvarapraṇidhāna is an important means to spiritual advancement in Patañjali's scheme. Many commentators as well as modern scholars tend to interpret *īśvara* as God and *īśvarapraṇidhāna* as devotion to God. Vyāsa (YB I.23; II.32) interprets it as a special kind of devotion (*bhaktiviśeṣa*), in which the devotee surrenders all his karmas (*sarvakarmārpaṇa*). But it is doubtful this is what Patañjali must have meant.

It will be clear from the forthcoming aphorisms that the notion of *īśvara* here is not that of a creator God, but rather of an ideal being who is absolutely free from defilements (*kleśa*), actions (*karma*), fruition of actions (*vipāka*) and latent impressions (*āśaya*). Hence, instead of translating the term *īśvara* as God, I have translated it as "the supreme being". The description suits the Buddha (especially seen in the Mahāyāna perspective).

The term *praṇidhāna* deserves special attention. Interpretation of the term as a special kind of devotion ignores the specific meaning the term has in Buddhist literature. Particularly in Mahāyāna Buddhism, *praṇidhāna* refers to an ardent desire, earnest wish or vow (Edgerton, p. 360). A common example of *praṇidhāna* is *bodhisattva*'s vow that he would become a Buddha. Hence, Vasubandhu refers to *bodhisattva-praṇidhāna* as follows: "This is how *bodhisattvas* perform *praṇidhāna*: In this blind world without a (spiritual) leader, I will be born as the Buddha, a master of those without a master".[44]

Against this background, *īśvarapraṇidhāna* can be consistently interpreted as one's desire or vow or resolve to become like *īśvara*. This would be consistent with Patañjali's conceptual framework, because *īśvara* for him is not the creator God, but rather an ideal being free from defilements, karma and so on. *Īśvarapraṇidhāna* in this sense could mean meditative concentration on *īśvara* with an intention to become *īśvara*, which in this context means to become an ideal being **like** *īśvara*.

क्लेशकर्मविपाकाशयैरपरामृष्टः पुरुषविशेषः ईश्वरः ।।24।।

Kleśakarmavipākāśayair aparāmṛṣṭaḥ puruṣaviśeṣaḥ īśvaraḥ.

The supreme being is a special kind of Self untouched by defilements, actions, their fruition and the latent impressions.

Comments

Here Patañjali is introducing the concept of *īśvara* in terms of the Sāṅkhya concept of *puruṣa* and the stereological concepts, namely *kleśa*, karma and *vipāka*. According to Sāṅkhya, *puruṣa*—the soul which has pure consciousness—is in fact beyond transmigration, but the characteristics of subtle body are attributed to the soul, due to which the categories of "bondage" and "liberation" happen to be attributed to it. Patañjali uses the soteriological categories of bondage (probably borrowing from Buddhism), namely defilements (*kleśa*), action (*karma*) and maturation (*vipāka*). (Vasubandhu refers to the Buddha's statement that there are three veils: the veil of action, the veil of defilement and the veil of maturation.)[45] To these three concepts, Patañjali seems to have added *āśaya*, which Vyāsa interprets as "appropriate latent impression" (*tadanuguṇavāsanā*). Here *āśaya* could be a modified form of *anuśaya*, or *āsava*.[46]

Patañjali is saying that *īśvara* is a special kind of *puruṣa* because he is untouched by defilements and so on. Other *puruṣa*s undergo bondage due to these binding factors (insofar as these factors are attributed to them), and they could be liberated when these binding factors are dissociated from *puruṣa* (that is, when the attribution of these binding factors stops). But in the case of the ideal *puruṣa*, this attribution itself does not take place. In this sense, *īśvara* is not a "liberated soul" (that is, the soul which was once in bondage but now is liberated); bondage simply does not apply to it. Vyāsa in his commentary explains this point.

This description suggests that Patañjali's *īśvara* could be understood as an ideal soul, the supreme being, but not as the God who creates the world.

तत्र निरतिशयं सर्वज्ञबीजम् ।।25।।

Tatra niratiśayaṁ sarvajñabījam.

In him there is the highest seed of omniscience.

Comments

Patañjali here describes *īśvara* as the extreme seed of omniscience. This simply seems to mean that *īśvara*, made of the nature of pure consciousness and untouched by ignorance and other defilements, can be the basis of omniscience.

It is worth noting that Patañjali is not saying that *īśvara* is omniscient. He is only saying that there is the seed of omniscience in him. This is probably because ordinary *puruṣa* can have knowledge through its association with *buddhi*. But *īśvara* is not associated with *buddhi*, because then he would be associated with karma also. Hence *īśvara* can be said to be the seed of omniscience, but not omniscient himself.

Vyāsa, however, interprets *īśvara* to be omniscient, as someone who has knowledge of everything and wants to impart knowledge out of compassion. He then identifies *īśvara* with Kapila (*ādividvān*: "the first knower"). who imparted knowledge to Āsuri. This interpretation is inconsistent with the basic concept of *īśvara* as *puruṣa* untouched by actions.

पूर्वेषामपि गुरुः कालेनानवच्छेदात् ।।26।।

Pūrveṣām api guruḥ kālenānavacchedāt.

Being unconditioned by time, he is the teacher of the ancients too.

Comments

Patañjali says that he is the teacher of the most ancient teachers because, being eternal, he existed before all earlier teachers. Patañjali does not explain in what sense *īśvara* can be a teacher. I think he cannot be called a teacher in the sense of one who teaches, by action or speech. But he can be a teacher because he is an ideal example which an aspirant wants to imitate through meditation.

तस्य वाचकः प्रणवः ।।27।।

Tasya vācakaḥ praṇavaḥ.

His designator (sign) is "*Om*".

Comments

Here the question which Patañjali seems to be addressing is how one can develop meditative concentration on *īśvara*. The question arises because *īśvara* as ideal *puruṣa* is formless, and it is not possible to concentrate on a formless object directly. One can do it through the medium of a tangible symbol.

In the case of *īśvara*, the symbol which designates it is "*Om*" (*praṇava*) according to Patañjali. Here Patañjali is influenced by *Upaniṣads*, according to which the ultimate reality such as *ātman* or *Brahman* is designated by *Om*.[47]

तज्जपस्तदर्थभावनम् ।।28।।

Tajjapas tadarthabhāvanam.

Its (*Om*'s) repetition is contemplation of its meaning

Comments

Here Patañjali means to say that repetition of the syllable *Om* should be done by a yogin not as a ritualistic practice, but with contemplation on the meaning of the syllable.

Vyāsa here identifies repetition of *Om* with *svādhyāya* (self-study), which may not have been intended by Patañjali. It seems that for Patañjali, repetition of *Om* with contemplation on meaning is part of the practice called *īśvarapraṇidhāna*. (As I understand it, it means resolve of the form "May I become like *īśvara*").[48] *Svādhyāya* (self-study) is a separate activity. Both activities are separately included under *kriyāyoga* (YS II.1) as well as under *niyama* (YS II.32).

ततः प्रत्यक्चेतनाधिगमोऽप्यन्तरायाभावश्च ।।29।।

Tataḥ pratyakcetanādhigamo'py antarāyābhāvaś ca.

From it results the understanding of individual Self and also the absence of obstacles.

Comments

Here Patañjali is stating the advantages of repetition of *Om* with contemplation on meaning. The advantages are two: (1) internal realization of consciousness and (2) removal of psychosomatic obstacles. The psychosomatic obstacles are stated in the next two aphorisms.

व्याधिस्त्यानसंशयप्रमादालस्याविरतिभ्रांतिदर्शनालब्धभूमिकत्वानवस्थितत्वानि चित्तविक्षेपास्तेऽन्तरायाः ।।30।।

Vyādhistyānasaṃśayapramādālasyāviratibhrāntidarśanālabdhabhūmikatvānavasthitatvāni cittavikṣepās te'ntarāyāḥ.

Disease, languor, doubt, heedlessness, sloth, indulgence, delusion, non-achievement of a stage and instability are the distractions of mind; they are obstacles.

Comments

Antarāya means obstacle. Here Patañjali is mainly concerned with psychological and psychosomatic conditions, which obscure the mind and obstruct the way leading to meditative trance or liberation. In this aphorism, Patañjali makes a synoptic and selective list of such obstructing conditions; in the next aphorism, he states some supplementary conditions. Most of these conditions were available in Buddhism under different heads. The heads under which these conditions can be found in Buddhism are *kleśa* (defilements), *upakleśa* (secondary defilements), *antarāya* (obstacles), *nīvaraṇa* (hindrance) and *saṁyojana* (fetters).

***Cittavikṣepa*(distractions of mind):** Patañjali uses the term *cittavikṣepa* as a common term for all obstacles. This has background in Abhidharma Buddhism. Vasubandhu refers to the view of Ābhidharmikas that "*cetaso vikṣepāḥ*" (mental distractions) are mental factors common to all defiled minds (*kliṣṭamahābhūmikāḥ*).[49] On this, Vasubandhu opines that since *samādhi* (concentration) is accepted as a factor common to all minds, distraction (*vikṣepa*) need not be mentioned separately because "when concentration itself is defiled it is called distraction".[50]

Antarāya : Patañjali's use of the term *antarāya* also has a background in Buddhism. Asaṅga, in *Śrāvakabhūmi* (ŚB: 144–9), discusses *antarāyas* in detail. He classifies *antarāyas* into internal and external (*ādhyātmam* and *bahirdhā*). Again, he classifies them into three kinds: obstacles in application of efforts (*prayogāntarāya*), obstacles in discriminative thinking (*prāvivekyāntarāya*) and obstacles in meditation or absorption (*pratisaṁlayanāntarāya*). He classifies the obstacles in meditation in two further types: obstacles in concentration meditation (*śamathāntarāya*) and obstacles in insight meditation (*vipaśyanāntarāya*).

Now let us consider the "obstacles" mentioned by Patañjali one by one and their possible Buddhist sources.

(1) ***Vyādhi***: *Vyādhi* (disease/illness) was mentioned by Asaṅga as an internal obstacle.[51] Under internal obstacles, Asaṅga includes intense mental passions and bodily diseases caused by the disorder of humours of body[52] and bad habits.
(2) ***Styāna***: Buddhists used the word *styāna* (Pali: *thīna*) in a wide sense which includes mental as well as bodily inactiveness. Vasubandhu defines *styāna* as "heaviness of body, heaviness of mind; inactiveness of body and inactiveness of mind".[53] Vasubandhu also refers to the Abhidharma view that *styāna* is of two kinds—bodily and mental.[54] Since Patañjali mentions *styāna* and *ālasya* separately, Vyāsa interprets *styāna* as mental inactivity and *ālasya* as bodily inactivity.
(3) ***Saṁśaya***: The Buddhist term for *saṁśaya* (doubt) is *vicikitsā*. *Vicikitsā* is included in many lists of defiled mental factors: it is one of the hindrances (*nīrvaraṇa*) as well as one of the fetters (*saṁyojana*). Vasubandhu includes "doubt about path" (*mārgasaṁśaya*) in the obstacles on the path to liberation.[55]
(4) ***Pramāda***: *Pramāda* (heedlessness) is one of the factors common to defiled minds.[56] Asaṅga includes *pramāda* among obstacles in concentration meditation (*śamathāntarāya*).[57]
(5) ***Ālasya***: Vyāsa defines *ālasya* (laziness) as "*kāyasya cittasya ca gurutvād apravṛttiḥ*" ("inactivity of body and mind due to heaviness"). As seen previously, the Buddhist concept of *styāna* includes bodily inactivity along with mental inactivity.
(6) ***Avirati***: *Avirati* literally means non-abstinence. The five precepts (or five restraints—*yamas*) are of the nature of abstinences (*viramaṇa* or *virati*). *Avirati* means opposite of that (non-abstinence). The term does not seem to be current in the Buddhist literature, though the content is very much there. Asaṅga refers to various acts of indulgence under the head *prāvivekyāntarāya* (obstacles in discrimination).[58]

The term *avirati* is prominently found in Jaina literature. According to Umāsvāti (the Jaina philosopher of the second century CE), there are five causes of bondage (*bandhahetu*): deluded world-view (*mithyādarśana*), non-abstinence (*avirati*), laxity (*pramāda*), passions (*kaṣāya*) and the actions of body, speech and mind (*yoga*).[59] It is quite possible that Patañjali's use of the word *avirati* is influenced here (like his discussion of *yama*)[60] by the Jaina account. Vyāsa explains *avirati* as craving (*gardha*) towards applying mind to sensuous objects.[61]

(7) ***Bhrāntidarśana*:** Literally means perceiving things with delusion. It is similar to Vasubandhu's idea of *mārgavibhrama* (AK V.44), that is, delusion regarding the path (which is one of the *mokṣāntarāyas*—obstacles to liberation). The idea is also similar to what Umāsvāti calls *mithyādarśana* (deluded worldview; see *avirati*).

(8) ***Alabdhabhūmikatva*:** Literally means not attaining a plane (*bhūmi*). In meditative practice, as per the Buddhist understanding, there are stages of meditation (for instance—four *dhyāna*s, then four *ārūpyas*) to which one is supposed to rise step by step. These stages are called *bhūmis*.

If because of other obstacles or insufficient practice one is not able to achieve the required plane, it will be called *alabdhabhūmikatva*. Though the idea is very much there in the Buddhist theory of meditation, I did not find this specific term in Buddhist literature. Patañjali might have coined it.

(9) ***Anavasthitatva*:** This means instability. It could mean unsteadiness or fickleness of mind. In Udānavarga, it is said that the wisdom does not become complete of the one who has an instable (*anavashita*) mind, does not understand the right way of life and has fickleness in confidence.[62] Asaṅga (ŚB: 148) talks about *cāpalya* (fickleness, unsteadiness) as an obstacle in Vipaśyanā. This notion of *cāpalya* has the same connotation as the term *anavasthitatva*. It is possible that Patañjali's notion of *anavasthitatva* is based on this background.

Vyāsa, however, explains this as not remaining stable in a plane (*bhūmi*) after attaining it.

On the whole, it can be said that Patañjali's notion of *antarāya/cittavikṣepa* is very much rooted in the Buddhist (and Jaina) discussion of the obstacles to emancipation.

दुखःदौर्मनस्याङ्गमेजयत्वश्वासप्रश्वासा विक्षेपसहभुवः ।।31।।

Duḥkhadaurmanasyāṅgamejayatvaśvāsapraśvāsā vikṣepasahabhuvaḥ

Pain, despair, shaking of limbs, (irregular) inhalation and exhalation arise along with these distractions.

Comments

Here Patañjali is referring to other accompanying signs or conditions of mental distraction. We come across direct or indirect references to them in Buddhist literature. Here four conditions are mentioned:

(1) ***Duḥkha*:** suffering or pain (which could be bodily or mental)
(2) ***Daurmanasya*:** Sadness (this is mental in nature)

(3) ***Aṅgamejaya*:** involuntary bodily movements/trembling
(4) ***Śvāsapraśvāsa*:** inbreathing and outbreathing

The terms *duḥkha* and *daurmanasya* many times occur together in Buddhist literature. Vasubandhu in AK (II.9) uses the term *duḥkhe* (meaning two *duḥkhas*) in the sense of *duḥkha* and *daurmanasya*. There the term *duḥkha* probably means physical pain or painful sensation, whereas *daurmanasya* means mental pain or despair. This pair (*duḥkha* and *daurmanasya*) is the negative counterpart of the other pair "*sukha* and *saumanasya* (often called *prīti*)".[63] Vasubandhu includes *duḥkha* and *daurmanasya* under extremely impure faculties (*ekānta-sāsrava indriyas*; AKB II.9cd.) He also says that they are associated with hatred.[64] Of course, Vasubandhu does not mention them as obstacles in meditation, as Patañjali does. But this could be said as implied.

There is no specific mention of trembling of body (*aṅgamejayatva*) as an obstacle in Buddhist literature, but a reference is found according to which the practice of mindfulness of breathing helps one in controlling the trembling.[65]

The conditions of *śvāsa* and *praśvāsa* need special attention. The literal meaning of *śvāsapraśvāsa* is "inbreathing and outbreathing". But Patañjali by this term does not mean either normal breathing or the special "breathing" which constitutes *prāṇāyāma*. By this term, he must be referring to defective breathing.

Asaṅga in *Śrāvakabhūmi* explains the defects (*apakṣāla*) possible in the practice of breathing and also how they are associated with mental distraction or mental imbalance. As he says:

> Two defects (are possible) in the practice of inbreathing and outbreathing (*āśvāsa-praśvāsa*): either congestion (*aśithilaprayogatā*) of breathing or (forcible) stoppage of breathing. Out of them, by congested breathing one becomes indolent; either mind gets covered by sloth and torpor or it gets distracted externally. By forcible stoppage of breathing, bodily imbalance or mental imbalance is generated. How is bodily imbalance generated? When one suppresses inbreathing and outbreathing by applying force, uneven gasses (winds) flow in the body. Initially those winds vibrate (*sphuranti*) in different parts of the body. They are called vibrators. These vibrating gasses, when they grow, cause pains in different parts of the body. This is called bodily imbalance. How is mental imbalance caused? His mind gets distracted (*cittaṁ vikṣipyate*) or is dominated by sadness (*daurmanasya*) and irritation (*upāyāsa*). Mental imbalance is generated in this way.
>
> (*ŚB: 222; translation mine*)

It is possible that this discussion of defective breathing practice and its association with mental distractions is at the background of the previous aphorism of Patañjali.

तत्प्रतिषेधार्थमेकतत्त्वाभ्यासः ।।32।।

Tatpratiṣedhārtham ekatattvābhyāsaḥ.
For their prevention the practice (of meditation) upon a single object (is prescribed).

Comments

Here Patañjali simply says that in order to remove or get rid of these mental distractions and their accompanying conditions, a yogin should practice meditation on a single object. Vyāsa, however, takes this as an opportunity to criticize the Buddhist doctrine of momentariness. Vyāsa says, "For the one, who holds that the mind is just the experience limited to its own object and is momentary, every mind is single-pointed and hence there is no distracted mind at all".[66] He argues that in order to make the possibility of "distracted mind" and "mind concentrated on a single object", we have to accept continuous identity of mind and the possibility of a single mind having many objects and choosing a single object for concentration.

The Buddhist who believes in momentariness would answer the objection by explaining the phenomenon in terms of minds as well as objects as forming different causal series, and "concentration on a single object" as meaning a consciousness series as corresponding to a homogenous object series and distraction as the one corresponding to a heterogenous object series.

मैत्रीकरुणामुदितोपेक्षाणां सुखदुःखपुण्यापुण्यविषयाणां भावनातश्चित्तप्रसादनम् ।।33।।

Maitrīkaruṇāmuditopekṣāṇāṃ sukhadukhapuṇyāpuṇyaviṣayāṇāṃ bhāvanātaścittaprasādanam.
The mind becomes tranquil by cultivation of friendliness, compassion, gladness and equanimity (or indifference), which have happiness, suffering, merit and demerit (respectively) as their objects.

Comments

Here Patañjali recommends the four sublime attitudes as the objects of meditative practice (*bhāvanā*), namely friendliness, compassion, gladness (appreciative joy) and equanimity (or indifference).[67] Vyāsa in his commentary interprets the four contemplations prescribed by Patañjali as follows:

> There one should practice friendliness towards all those beings who happen to have pleasure and enjoyment. He should practice compassion towards suffering beings. He should practice joy towards meritorious beings and indifference (*upekṣā*) towards those having sinful character. When he practices in this way, there arises white character (*śuklo dharmaḥ*, that is, pure character). By that the mind becomes tranquil. A tranquil one-pointed mind attains a stable state.
>
> *(YB I.33; translation mine)*

I suggest here that Patañjali's approach to the four attitudes has to be understood differently. For example, when Patañjali says that *maitrī* has happiness as its object, it simply means that developing *maitrī* implies desiring for happiness of the other. Similarly, when Patañjali claims that *karuṇā* has pain as its object, it may simply mean that developing *karuṇā* about another person implies desiring that the pain of the other person may be removed. Similarly, to say that *muditā* has merit as its object implies that developing *muditā* towards another person implies appreciating the merit of the other person, and to say that *upekṣā* has demerit as its object implies that one should be detached from the demerit in the other person.

But if this is the way happiness, suffering, merit and sinfulness can be the objects (*viṣaya*) of friendliness, compassion, appreciative gladness and indifference (or detachment) respectively, then these attitudes may not be directed to specific types of persons, but they could be directed to all. Hence, *maitrī* will not be directed to happy persons only; *maitrī* as a wish for another person's happiness would be directed towards all. Similarly, the other three attitudes can be addressed to all beings in the world. This gives us what can be called a universalist interpretation of the four attitudes.

In this context, the distinction between the universalist formulations and specificist formulations of the four attitudes becomes relevant. In fact, we find both the formulation in Indian tradition and the question as to Patañjali's approach to the four attitudes may be investigated against the background of different formulations of the four attitudes we find in the diverse Indian tradition.

(1) Universalist formulation

According to this formulation, the four attitudes (namely *maitrī*, *karuṇā*, *muditā* and *upekṣā*) are addressed to all beings. In Buddhism, we generally get this formulation. In fact, the four attitudes were called *apramāṇa* (*appamaññā* in Pali) in Buddhist literature, meaning "immeasurable", for the same reason. Asaṅga in *Abhidharmasamuccaya* (ADS: 94–95) formulates the four attitudes as:

> ***Maitrī***: *sattvāḥ sukhena saṁyujyeran* (May beings be associated with happiness).
> ***Karuṇā***: *sattvāḥ duḥkhena viyujyeran* (May beings be dissociated from suffering).
> ***Muditā***: *sattvāḥ sukhena na viyujyeran* (May beings not be dissociated from pleasure).
> ***Upekṣā***: *sattvā hitaṁ labheran* (May beings attain good).

Other Buddhist scholars formulated these attitudes differently. There is not much diversity in the formulation of *maitrī* and *karuṇā*. But there is greater diversity in formulation of *muditā* and *upekṣā*. See for instance Vasubandhu's formulation:

> ***Maitrī***: May all beings be happy (*sukhitā bata*).
> ***Karuṇā***: Oh! All beings are unhappy! (*duḥkhitā bata*).

***Muditā*:** May all beings rejoice (*modantām*).
***Upekṣā*:** These are living beings (*sattvāḥ*).[68]

Here *upekṣā* implies realizing beings just as beings (with indifference towards their happiness or unhappiness).

(2) Specificist formulation

Against this, we get specificist formulations in non-Buddhist traditions. By specificist formulations, I mean where different attitudes are recommended to be developed towards specific types of beings and not towards all. In the *Carakasaṁhitā*, for instance, the four attitudes are recommended as follows:

> A physician should have fourfold attitude: *maitrī* and *karuṇā* with sick persons, joy with regard to curable beings and indifference/avoidance (*upekṣā*) with reference to those nearing death.[69]

Similarly, in the Jaina classic *Tattvārthasūtra*, the four attitudes are recommended in a specificist way as follows:

> The observer of vows should cultivate friendliness towards all living beings, delight (*prāmodya*) towards those who exceed in merit, compassion (*kāruṇya*) towards miserable ones and equanimity towards the vainglorious.[70]

Now the question is as to what the formulation of the four sublime attitudes intended by Patañjali could be. Could it be universalist or specificist?

Here I want to argue that Vyāsa has not correctly interpreted Patañjali. Patañjali says that *maitrī*, *karuṇā*, *muditā* and *upekṣā* have *sukha*, *duḥkha*, *puṇya* and *apuṇya* as their respective objects. If Patañjali wanted to say the way Vyāsa interpreted it, then he could have said that *maitrī*, *karuṇā muditā* and *upekṣā* are about *sukhī*, *duḥkī*, *puṇyavān* and *apuṇyavān*. So Patañjali must mean something else. When Patañjali says that *maitrī* has *sukha* as its *viṣaya* and *karuṇā* has *duḥkha* as its *viṣaya*, it simply means that *maitrī* is about happiness and *karuṇā* is about pain. It does not mean that *maitrī* should be addressed to happy persons and *karuṇā* to unhappy persons. This is clear in Asaṅga's formulation, which we have already seen. By following Asaṅga's formulation as the model, we can construct formulations of the four sublime attitudes suggested by Patañjali's aphorism as follows:

***Maitrī*:** May all beings be associated with happiness (*sarve sattvāḥ sukhena saṁyujyeran*).
***Karuṇā*:** May all beings be dissociated from sufferings (*sarve sattvā duḥkhena viyujyeran*).

Muditā: May all beings be associated with merit (*sarve sattvāḥ puṇyena saṁyujyeran*).
Upekṣā: May all beings be dissociated from demerit (*sarve sattvā apuṇyena viyujyeran*).

This interpretation does better justice to Patañjali's own formulation. It implies that Patañjali's conception of the four sublime attitudes must have been universalistic (and not specificist) and it must have been influenced by the Buddhist approach to them. Vyāsa, however, interprets it (wrongly) as specificist. This explanation leaves open the question about *upekṣā*. The interpretation of *upekṣā* suggested by this formulation which I have attributed to Patañjali is different from generally accepted interpretations of it. There are two generally accepted interpretations of *upekṣā*:

(1) One is its literal meaning, such as indifference or avoidance. For example, when Caraka says that a physician shall have *upekṣā* towards the patients nearing death, he expresses this meaning. Similarly, Vyāsa too interprets Patañjali's notion of *upekṣā* as the attitude of indifference towards sinful persons. This can be regarded as the negative meaning of *upekṣā*.
(2) But the word *upekṣā* also has a more constructive meaning, namely *samatva* or *mādhyasthya*, that is, equanimity. In their theory of meditation, Buddhists generally use the term *upekṣā* in this sense. *Upekṣā* in this sense can be an object of meditative practice.[71]

Patañjali's conception of *upekṣā*, which I have formulated as "May all beings be dissociated from sin (*apuṇya*)", is different from both. But it is not completely exceptional, because we have seen that Asaṅga formulated *upekṣā* as "May all beings be associated with good (*hita*)". It is not too different from Patañjali's formulation as I have suggested.[72]

Vyāsa's formulation of *upekṣā* is inconsistent with Patañjali's approach for another reason. Patañjali in this aphorism recommends meditative practice (*bhāvanā*) of all four attitudes. But if *upekṣā* simply means indifference towards a sinful person or avoidance of a sinful person, then it cannot be meditated upon. Vyāsa does not realize this problem while commenting on this aphorism. He realizes it while commenting upon YS III.23. He says there that *bhāvanā* is possible with respect to other sublime attitudes, but *upekṣā* of the persons of sinful character cannot have the status of *bhāvanā*.[73] Vyāsa does not realize that this problem arises because of a specificist interpretation of the four attitudes, and that universalistic interpretation does not have this problem.

प्रच्छर्दनविधारणाभ्यां वा प्राणस्य ।।34।।

Pracchardanavidhāraṇābhyāṃ vā prāṇasya.

Or (the mind becomes tranquil) by expulsion and retention of breath.

Comments

Here Patañjali seems to be referring to *prāṇāyāma* in Haṭhayogic sense. He is saying that tranquillity of mind can be attained by complete exhalation (*prac-chardana*, which literally means vomiting) or by holding the breath. Though we do not find Haṭhayoga texts at the time of Patañjali, there is no doubt that Haṭhayoga practices were prevalent. One of the basic principles of Haṭhayoga was that operations of mind and breathing are causally interrelated. So Patañjali seems to mean here that by the technique of complete exhalation we can empty the mind of thoughts and by holding the breath we can make the mind stable.

However, what Patañjali calls *prāṇāyāma*, as one of the limbs in the eight-limbed Yoga, is closer to the Buddhist technique of mindfulness of breathing, as it will be argued there (YS II.49–51).

विषयवती वा प्रवृत्तिरुत्पन्ना मनसः स्थितिनिबन्धिनी ।।35।।

Viṣayavatī vā pravṛttir utpannā manasaḥ sthitinibandhinī.

Or an attitude (of sense organ or mind) arising along with an object, (when concentrated upon,) conduces to mental steadiness.

Comments

When a sensory cognition or a mental state along with its object arises, one can observe it as it is. This causes stability of mind.

This can be compared with *cittānupassanā* as a part of *satipaṭṭhāna* (mindfulness meditation). *Cittānupassanā* consists of mindful observation of one's present state of mind. Asaṅga in *Śrāvakabhūmi* (pp. 294–9) discusses different states of mind which the meditator is supposed to observe from time to time.

Of course, the difference remains that the method of *cittānupassanā* has been discussed in Buddhism as a part of mindfulness meditation, whereas Patañjali is prescribing it as a part of concentration meditation.

Vyāsa interprets "*viṣayavatī pravṛttiḥ*" as "the activity of a sense-organ towards a special object (such as divine odour)".[74] This may not be intended by Patañjali.

विशोका वा ज्योतिष्मती ।।36।।

Viśokā vā jyotiṣmatī.

Or the (mental attitude which is) sorrowless and full of luminosity (conduces to mental steadiness).

Comments

Here Patañjali is referring to a special state of mind as the object of concentration—the state of mind which is happy (literally: without sorrow—(*viśokā*) and full of luminosity (*jyotiṣmatī*)). It is possible that this is Patañjali's own proposal. The exact source is unknown. Vyāsa tries to explain it in his own way.[75]

वीतरागविषयं वा चित्तम् ।।37।।

Vītarāgaviṣayaṃ vā cittam.

Or the mind having an Attachment-free person for its object (of meditation) (becomes steady).

Comments

Vītarāga means a person free from attachment. He is an ideal person according to Buddhist soteriology. He is the ideal person among those following the mundane path to liberation (*laukikamārga*).[76] (*Arhat* is supposed to be the ideal of the extra-mundane path (*lokottaramārga*)). Asaṅga describes *vītarāga* as follows:

> What are the signs of an attachment-free person (*vītarāga*)? One can say: His bodily actions are stable, sense-organs are firm. His manner of behaviour (*īryāpatha*) cannot be surpassed easily. He spends a long time with a single manner of behaviour without getting disturbed. He does not desire to shift to a different manner of behaviour. He speaks slowly, speaks peacefully. He does not take pleasure in mixing with the crowd; does not take pleasure in contact (with sensuous objects). His speech operates steadily. Having seen forms with eyes he entertains the forms; does not entertain attachment to the forms. In the same way he entertains sound, odour, taste and touch, but does not entertain attachment to them. He is fearless; has thorough intellect, with mind and body containing great confidence. He is not covetous and cannot be agitated. He is merciful and the vicious thoughts of sensuous desire do not disturb his mind.
>
> (*ŚB: 469–70; translation mine*)

However, though *vītarāga* is an important concept in Asaṅga's scheme, it is supposed to be the ideal to be reached by following a moral–spiritual path at the mundane level. It is not made an object of meditation. Patañjali makes *vītarāga* an object of meditation. The idea of contemplation on *vītarāga* is comparable to the idea of *anusmṛti* (*anussati*, in Pali: contemplation/reflection) found in the Theravāda theory of *kammaṭṭhāna* ("objects of meditation"),[77] which includes contemplation on the Buddha, on deities and so on.[78] It is not known whether this idea would have been known to Patañjali.

स्वप्ननिद्राज्ञानालम्बनं वा ।।38।।

Svapnanidrājñānālambanaṃ vā.

Or the mind which takes the cognition of dream or sleep for its object (of meditation) (becomes steady).

Comments

It is important to develop the correct awareness about dream experience when one remembers one's own dream—that it was only a dream experience and not a waking experience. But it is difficult to understand how concentration on dream experience can cause stability of mind. Similarly, it is important to be mindful or alert before going to sleep, after sleep or in the intervals between two sessions of sleep.[79] But it is difficult to appreciate the idea that concentration on sleep-experience can lead to stability of mind.

यथाभिमतध्यानाद्वा ।।39।।

Yathābhimatadhyānād vā.

Or (mind becomes steady) by meditation of any object according to one's inclination.

Comments

The first chapter of the *Yogasūtra* is significant for its prescription of diverse objects and techniques as means to achievement of concentration or tranquillity of mind or meditative absorption. Patañjali in YS I.17 starts with *samprajñāta-samādhi*. In I.21–22, he says that one gets results according to the intensity of one's fervour (*saṁvega*). He goes on to prescribe other objects of meditation—such as *īśvara*, the four *bhāvanās* (*maitrī*, *karuṇā*, *muditā* and *upekṣā*), breathing, mindfully observing mental states or concentrating on an ideal person (*vītarāga*). In this aphorism, Patañjali concludes by taking the most liberal position. He is saying here that the mind can become stable by concentrating the object of one's own inclination.

We also find prescription of diverse objects in the Buddhist theory of meditation. In Theravāda, which developed in Sri Lanka, we have the doctrine of forty *karmasthānas* (objects of meditation). They include ten devices called *kasiṇas* (Sanskrit *kṛtsnas*: holistic forms), ten impurities (*aśubhas*), ten contemplations (*anusmṛtis*), four illimitables or modes of sublime conduct (*brahmavihāras*), one perception (*saṁjñā*: perception of loathsomeness of material food), one analysis (analysis of four elements in the body) and four formless spheres.[80]

The theories of Asaṅga and Vasubandhu which seem to be available to Patañjali do not contain all forty objects of meditation, though they do contain a number of them. In Asaṅga, for instance, we find the mention of *kṛtsnas*, *aśubhas*, formless spheres and *brahmavihāras* and also some *anusmṛtis* (such as *ānāpānasmṛti*).

Some objects prescribed by Asaṅga and Vasubandhu are included in Patañjali's list: *vitarka*, *vicāra*, *ānanda* (*prīti* and *sukha* of Buddhist meditation), breathing, the four *bhāvanās* (*brahmavihāras*) and mental states. Patañjali omits some. In particular he omits *aśubhas* (impurities) as the objects of meditation. But he adds some objects, such as *vītarāga*, dream-experience, sleep and *īśvara*.

Just as Buddhists were liberal in prescribing diverse objects, Patañjali is also liberal, as we see in this aphorism. But there is important difference between the two liberal approaches. Buddhist prescription of diverse objects is regulated by their doctrine of temperaments. The specific objects are prescribed to the aspirants according to their temperament. Hence Asaṅga says in *Śrāvakabhūmi*:

> What is "propriety in application (of mind)" (*anurūpa-prayogatā*)? If one has attached temperament (*rāgacarita*) then one fixes mind on impurity. If one has hateful temperament (*dveṣacarita*), then one focuses on loving kindness and so on. Ultimately, if one has thinking temperament (*vitarka-carita*), then one focuses mind on the mindfulness of breathing. But a person with equanimous conduct (*samabhāga-carita*) and weak passion (*mandarajaska*) applies mind to the object in which he develops liking (*yatrālambane priyārohatā bhavati*). This is called propriety in application (of mind).
>
> *(ŚB: 389; translation mine)*

Hence Asaṅga also prescribes "any object one likes", but he prescribes it not for all, only to a person with equanimous conduct and weak passion.

The term "*yathābhimata*", used by Patañjali in the previous aphorism, can also be taken to mean, "according to what one has accepted/admitted",[81] in which case Patañjali will appear as having an open or liberal approach towards different sectarian beliefs (including Buddhism and Jainism). In other words, although Patañjali seems to accept the Sāṅkhya metaphysical vision as ultimate, he seems to be devising a technique (based on Upaniṣadic, Buddhist and Jaina material) which could be commonly available to persons of different sects. The openness or liberality of his approach is also suggested by his frequent use of the word *vā* ("or"), suggesting the availability of alternative means.[82] Such a pluralistic approach is suggested by some other aphorisms as well, such as the aphorisms on self-study.[83] However, later commentators of YS tended to present Patañjali as a sectarian thinker.

परमाणुपरममहत्त्वान्तोऽस्य वशीकारः ।।40।।

Parmāṇuparamamahattvānto'sya vaśīkāraḥ.

Then its (=his mind's) mastery extends from the smallest thing to the largest magnitude.

Patañjali is stating here a general advantage of meditative concentration. He is claiming that through such a meditative concentration on different objects, one attains mastery over all types of objects, ranging from the subtlest to the largest ones.

The idea that one can achieve mastery over objects through deep meditation on them has been emphasized by Buddhists in the context of supernormal powers.[84] In Buddhism, a close connection between concentration on whole material objects (*kasiṇa* in Pali, *kṛtsnāyatana* in Sanskrit) and mastery over the objects was acknowledged.[85] According to Vasubandhu, through concentration on the "whole objects", one enters into mastery (*abhibhvāyatana*). Through mastery, one enters into freedom (*vimokṣa*).[86]

In his commentary, Vyāsa says that by concentrating on subtle objects, one's mind attains stability down to atomic objects, and by concentrating on gross objects it attains stability up to extremely large objects. The yogin then experiences non-obstruction or free flow, and this itself is mastery (*vaśīkāra*). This interpretation of subtle and gross is closer to Vaiśeṣika understanding rather than Sāṅkhya understanding. Similarly, Patañjali's use of the words *paramāṇu* and *paramamahat* refer to the extreme dimensions acknowledged by Vaiśeṣikas. This suggests that Patañjali and Vyāsa adopt Vaiśeṣika ontology when they find it suitable.[87]

क्षीणवृत्तेरभिजातस्येव मणेर्ग्रहीतृग्रहणग्राह्येषु तत्स्थतदञ्जनता समापत्तिः ।।41।।
Kṣīṇavṛtter abhijātasyeva maṇer grahītṛgrahaṇagrāhyeṣu tatsthatadañjanatā samāpattiḥ.
When the modifications of mind are reduced, mind becomes like a transparent crystal; the state of such mind with respect to the grasper, grasping or graspable and its act of assuming their shape is called *samāpatti* (meditative attainment).

Comments

Here Patañjali is introducing the notion of *samāpatti*. According to him, it is a meditative process by which the mind, through concentration, becomes as if like the object of concentration, but remains unattached to it. In order to explain this dual role of mind ("closely associated, but unattached"), Patañjali uses the metaphor of a crystal which is as if coloured by the object with which it is associated, but in reality remains un coloured. Here Patañjali seems to mean "becoming" ("turning into", "getting transformed into") by the word *samāpatti*. His use of the expression "*tatstha-tadañjanatā*" suggests the state of mind in which the mind is associated with the object (*tatstha*) and becomes coloured by it (*tadañjana*). Of course, this "object", according to him, could be of the nature of grasper (*grahītṛ*: cognizer), grasping (*grahaṇa*: cognition) or the graspable (*grāhya*: the object of cognition).

Samāpatti is an important concept of the Buddhist theory of meditation. The term occurs frequently in the works of Asaṅga and Vasubandhu.[88] In Pali literature on meditation, the concept of *samāpatti* occurs in the form of the past participle (*samāpanna*) of the same root (*sam+ā+√pad*). *Samāpanna* means reached, attained, accomplished.[89] Hence, *samāpatti* means attainment, accomplishment. Here *samāpatti* is generally the attainment of *dhyāna* and hence, corresponding to every *dhyāna*, there can be attainment/achievement/accomplishment of that *dhyāna*. Thus, there is substantially no difference between *dhyāna* and *samāpatti*. That is why Vasubandhu says, "*samāpatti* means the wholesome (*śubha*) one-pointedness of mind"[90] and comments that "*samāpatti* is not different from meditation, that is, one-pointedness of wholesome mind, because it is meditative-concentration (*samādhi*) by its very nature".[91]

Patañjali's conceptualization of *samāpatti* differs slightly from the Buddhist conceptualization. According to the former, it means meditative identification of mind with the object (or meditative attainment of the object by mind). According to the latter, it is attainment of the state of meditation by the mind. (I am translating the term *samāpatti* in the previous aphorism as "meditative attainment" in order to cover both the senses, though not in a fully satisfactory way.)

The difference between the two connotations, as I have suggested, is not substantial, and it does not have any further implications. It is safe to conclude that Patañjali's concept of *samāpatti* might have been a reconstruction of the Buddhist concept.

In spite of the slight difference mentioned here, the further operations of the term *samāpatti* in both the systems are parallel. Just as *samādhi* (*rūpadhyāna*, according to Buddhists, or *samprajñāta-samādhi*, according to Patañjali) can be with or without *vitarka* or with or without *vicāra*, similarly *samāpatti*, according to both, can be *savitarka*, *nirvitarka*, *savicāra* and *nirvicāra*.

तत्र शब्दार्थज्ञानविकल्पैः संकीर्णा सवितर्का समापत्तिः ।।42।।

Tatra śabdādrthajñānavikalpaiḥ saṃkīrṇā savitarkā samāpattiḥ.

The meditative attainment (*samāpatti*) associated with a gross object is the one which is mixed up with words, meanings, cognitions and ideas.

Comments

Here Patañjali appears to provide a four-fold classification of *samāpatti*.

(1) *Savitarka-samāpatti*: meditative attainment with gross thought.
(2) *Nirvitarka-samāpatti*: meditative attainment without gross thought.
(3) *Savicāra-samāpatti*: meditative attainment with subtle thought.
(4) *Nirvicāra-samāpatti*: meditative attainment without subtle thought.

Numbers (1) and (2) refer to association or non-association with gross object (*sthūlaviṣaya*) and (3) and (4) refer to association with subtle object (*sūkṣmaviṣaya*). This four-fold classification could be misleading, because when we are dealing with "*samādhi* with object" (*samprajñāta-samādhi*), it then follows that if there is no gross object, a subtle object has to be there. As a result. it boils down to a three-fold classification:

(1) *Savitarka-savicāra*: *samāpatti* associated with gross object.
(2) *Nirvitarka-savicāra-samāpatti*: *samāpatti* without gross object but with a subtle object.
(3) *Nirvicāra-samāpatti*: *samāpatti* without even subtle object.

This will correspond to the first *dhyāna*, intermediate *dhyāna* (*dhyānāntara*) and second *dhyāna* of Buddhism.

Patañjali in this aphorism is defining *savitarka-samāpatti* as the one mixed with words, meanings, cognitions and mental construction. We do not have this definition in Buddhist literature. But one important point of similarity is that Vasubandhu acknowledges a close connection between *vitarka* and *vikalpa*. He remarks that the five consciousnesses which are associated with the gross form as well as the subtle form (*savitarka-savicāra*) are not devoid of natural mental construction.[92] According to him, natural mental construction (*svabhāva-vikalpa*) is the same as *vitarka*.[93]

स्मृतिपरिशुद्धौ स्वरूपशून्येवार्थमात्रनिर्भासा निर्वितर्का ।।43।।

Smṛtipariśuddhau svarūpaśūnyevārthamātranirbhāsā nirvitarkā.

The meditative attainment is called free-from-gross-objects, when one's memory is purified and when the concentration loses its own (subjective) nature, and shines forth in the form of object.

Comments

In the previous aphorism, Patañjali introduced "meditative attainment with gross thought". In this aphorism, he introduces a higher stage of meditative attainment, namely that without gross thought. Patañjali further claims that one can attain this higher stage when there is purification of *smṛti*. We have seen while discussing *smṛti* as a *citta-vṛtti* according to Patañjali that though there Patañjali is referring to "memory" by the word *smṛti*, his definition of *smṛti* as "non-removal of the experienced object" seems to be inspired by the Buddhist definition of mindfulness.[94] So here it is a question whether Patañjali means by *smṛti-pariśuddhi* purification of memory or that of mindfulness.

The first possibility (purification of memory) can be explained as follows. A gross thought, whether moral or immoral, is a heterogenous and disturbing thought. It becomes disturbing because of passions. Memory plays an important part in the disturbing role of passions. Thought of one object leads to recollection of another object to which we are emotionally attached, and in this way the thought becomes more complex and disturbing. Purification of memory is closely associated with reduction of gross (disturbing) thoughts.

In the Buddhist theory of meditation, however, the term *smṛti-pariśuddhi* has a special significance and means "purification of mindfulness". According to the Buddhist theory, *smṛti-pariśuddhi* is one of the factors in the fourth *dhyāna*.[95] Vasubandhu describes the fourth *dhyāna* as "the last one. In that there are four factors: the neither-pleasant-nor-painful sensation, purification of indifference (or equanimity), purification of mindfulness and concentration".[96] The idea is that in the third *dhyāna* equanimity and mindfulness arise as the additional factors, and they are coupled with pleasant sensation (*sukha*). *Sukha* drops out of them in the fourth *dhyāna*, and though equanimity and mindfulness remain, they are purified. *Vitarka* and *vicāra* (gross and subtle thoughts) have ceased in the second *dhyāna* itself.

It is quite probable that Patañjali is inspired by the Buddhist concept of *smṛti-pariśuddhi*, as explained, when he says that the meditative attainment becomes devoid of gross thoughts when his mindfulness is purified. However, here Patañjali seems to be altering the causal sequence found in the Buddhist theory. The

Buddhists say this the other way: "When gross and subtle thoughts cease (in the second *dhyāna*), then subsequently mindfulness gets purified (in the fourth *dhyāna*)".

Patañjali's definition of *nirvitarka-samāpatti* has another dimension. It involves "concentration, which as if loses its own nature, and manifests the object alone"; this is the feature shared by *samādhi* as defined in YS III.3. This suggests a parallelism between the theory of *savitarka-nirvitarka-savicāra-nirvicāra-samāpatti*, on the one hand, and that of *dhāraṇā-dhyāna-samādhi*, on the other.[97]

एतयैव सविचारा निर्विचारा च सूक्ष्मविषया व्याख्याता ।।44।।

Etayaiva savicārā nirvicārā ca sūkṣmaviṣayā vyākhyātā.

By this, the concentrations having subtler objects, viz., "the one accompanied by subtle thought", and "the one free from subtle thought" have been explained.

Comments

Here Patañjali is saying that meditative attainments with and without subtle thought can be explained on the same lines as those with and without gross thought. That the distinction between *vitarka* and *vicāra* is between gross and subtle is the view found in Vasubandhu, when he says "grossness of mind is *vitarka* and subtleness of mind is *vicāra*".[98] This suggests that in the *vitarka* stage the mind (*citta*) itself becomes gross (*udāra=sthūla*) and in the *vicāra* stage, the mind itself becomes subtle. Patañjali, on the other hand, suggests that grossness and subtleness in *vitarka* and *vicāra* respectively are the characteristics of objects of thought. (However, there is no absolute clarity on this issue either in Buddhism or in Patañjali's Yoga.)

By equating the *savitarka-nirvitarka* division with the *savicāra-nirvicāra* division, Patañjali suggests that though *vicāra* is subtle as compared to *vitarka*, a higher meditative attainment should be devoid of both because the difference between them is of degree rather than of kind. Secondly, purification of mindfulness has a role to play in reaching the higher level of meditative attainment in both cases.

It should be noted here that the two divisions are not parallel as they appear to be but rather are overlapping divisions. One and the same meditative attainment can, for example, be without gross thought but with subtle thought.

सूक्ष्मविषयत्वं चालिङ्गपर्यवसानम् ।।45।।

Sukṣmaviṣayatvaṃ cāliṅgaparyavasānaṃ.

And the province of the (concentration with) subtle objects reaches up to the non-mark (that is, the primordial nature).

Comments

As pointed out in the comments under YS I.44, *vitarka* and *vicāra*, according to Vasubandhu, refer to grossness and subtleness of mind itself, whereas grossness and subtleness in *vitarka* and *vicāra* respectively are the characteristics of objects of thought, according to Patañjali. Now it becomes possible for Patañjali to explain grossness and subtleness of objects as they are defined in the classical Sāṅkhya. Sāṅkhyas are known for their model of cause–effect relation, where the effect is the manifest or gross form assumed by the cause, so that the cause can be inferred on the basis of the effect. In this sense, the effect is the mark (that is, inferential mark: *liṅga*) of the cause. According to the Sāṅkhya theory of creation, the five gross elements are the marks (*liṅga*) of five elemental essences (*tanmātras*); the five elemental essences are the marks of the ego (*ahaṁkāra*); the ego is the mark of the cognitive faculty (*buddhi*); and that in its turn is the mark (what Vyāsa calls "mark alone": *liṅgamātra*) of the primordial nature (*pradhāna*). But the primordial nature is not the manifestation of anything. Hence, Patañjali calls it the non-mark (*aliṅga*).

ता एव सबीजः समाधिः ।।46।।

Tā eva sabījaḥ samādhiḥ.

These are the seeded concentrations.

Comments

After introducing *savitarka, nirvitarka, savicāra* and *nirvicāra-samāpatti*, Patañjali identifies them here as *sabīja-samādhi* (*samādhi* with seed). Vyāsa interprets *sabīja-samādhi* as the *samādhi* having an external thing as the seed (*bahirvastu-bīja*). Vyāsa's interpretation is problematic.

First of all, *samāpatti*, according to Patañjali, is not necessarily about an external object; it can be about an internal object as well. According to his definition (YS I.41), *samāpatti* can be about the grasper (*grahītṛ*), grasping (*grahaṇa*) or the object of grasping (*grāhya*). Hence, the interpretation of *bīja* as an "external thing" (*bahirvastu*) seems to be untenable.

Patañjali himself seems to use the term *bīja* in a different way. From the aphorisms YS I.50 and YS I.51, it is clear that the meditative absorption becomes seedless (*nirbīja*) due to the cessation of *saṁskāras*. Hence, *bīja* should be taken to mean *saṁskāra*. This implies that Vyāsa's identification of *samprajñāta-samādhi* with *sabīja-samādhi* and *asamprajñāta-samādhi* with *nirbīja-samādhi* (as he does in YB I.2) may not be correct. It is clear from Patañjali's definition that *asamprajñāta-samādhi* possesses *saṁskāra*.[99]

Hence *sabīja-samādhi* should be understood as *samādhi* with *saṁskāras* (that is, latent impressions) such that it includes both *samprajñāta* and *asamprajñāta-samādhi*. Identification of *bīja* with *saṁskāras* seems to be the reflection of the view of early Yogācāra, which identifies *bīja* with *ālayavijñāna*. Here *ālayavijñāna* is the collection of the latent impressions (*vāsanās*, which Patañjali calls *saṁskāras*).[100]

निर्विचारवैशारद्येऽध्यात्मप्रसादः ।।47।।

Nirvicāravaiśāradye'dhyātmaprasādaḥ.

Confidence in the meditative attainment, free-from-subtle-thoughts, results into internal calmness.

Comments

Vaiśāradya is a Buddhist technical term which means fearlessness or confidence.[101] Hence, *nirvicāravaiśāradya* would mean fearlessness or confidence that one obtains by the third *dhyāna* (which is devoid of *vitarka* and *vicāra*).

Patañjali's term *adhyātmaprasāda* seems to be an abridged form of the Buddhist term *adhyātmasamprasāda*, which means internal tranquillity (that is, tranquillity that one feels inside). According to the Buddhist theory of meditation, this is a factor in the second *dhyāna*.[102] This tranquillity arises due to the cessation of the turbulence caused by gross and subtle thoughts (*vitarka* and *vicāra*). Vasubandhu explains:

> What is this phenomenon called internal tranquility (*adhyātmasamprasāda*)? Internal tranquility is the peaceful flow of the mind-series due to the absence of the turbulance of gross and subtle thoughts. The mind series turbulated by gross and subtle thoughts is disturbed (not tranquil) like a river having tides.[103]

In non-Buddhist tradition, the word *vaiśāradya* refers to proficiency or serenity. Accordingly, Vyāsa takes *vaiśāradya* to mean clean flow of mind and interprets cleanliness in terms of removal of *rajas* and *tamas*. However, it is highly likely that Patañjali is using the term with the Buddhist connotation rather than the one current in non-Buddhist tradition.

In his commentary, Vyāsa quotes a verse which seems to be derived from Buddhist sources.[104]

ऋतंभरा तत्र प्रज्ञा ।।48।।

Ṛtaṃbharā tatra prajñā.

In that stage (arises) the truth-bearing wisdom.

Comments

In YS I.47, Patañjali referred to "internal tranquility" (*adhyātmaprasāda*) as a result of confidence generated by *nirvicāra* meditative attainment. In this aphorism, he refers to *prajñā* (wisdom, insight) as a result of that. This is consistent with the Buddhist theory of meditation, according to which the second *dhyāna* is without *vicāra* but is accompanied by *adhyātmasamprasāda*[105] and the third *dhyāna* is accompanied by equanimity, mindfulness, wisdom (or insight), bliss and concentration.[106]

In Pali tradition, the word used for wisdom (or insight) in the third *dhyāna* is *sampajañña*, the Sanskrit analogue of which is *samprajñāna*. Interestingly, in his formulation of the theory Vasubandhu uses the word *prajñā* in the verse AK VIII.8 and interprets it as *samprajñāna* in the commentary.[107]

The term *ṛtambharā prajñā* (truth-bearing wisdom/insight) seems to have been coined by Patañjali on the basis of the Vedic concept of *ṛta* (truth). The term means "the insight into the true nature of things". According to Tandon, it refers to the "*bhāvanāmayā paññā*"[108] (wisdom based on meditation), which is the highest wisdom in the three-fold classification of wisdom. The three-fold classification of wisdom will be discussed under the next aphorism.

श्रुतानुमानप्रज्ञाभ्यामन्यविषया विशेषार्थत्वात् ।।49।।

Śrutānumānaprajñābhyām anyaviṣayā viśeṣārthatvāt.

It (that is, the truth-bearing wisdom) has objects different from those of wisdom based on verbal testimony and on inference, because it has particulars as its objects.

Comments

Here Patañjali is distinguishing truth-bearing wisdom (*ṛtambharā prajñā*) from two other types of wisdom (*prajñā*): wisdom based on verbal testimony (*śruta-prajñā*) and that on inference (*anumāna-prajñā*). This presupposes a three-fold classification of wisdom. The three-fold classification of *prajñā* was available in Buddhism: *śrutamayī*, *cintāmayī* and *bhāvanāmayī*.[109] *Śrutamayī prajñā* is the wisdom based on what one hears or what one knows from scriptures. *Cintāmayī prajñā* is the wisdom based on thinking or reasoning. And *bhāvanāmayī prajñā* is the wisdom (or understanding, insight) based on meditation. Patañjali's three-fold distinction seems to be clearly based on the Buddhist distinction. Hence, Patañjali's *anumāna-prajñā* is derived from Buddhism's *cintāmayī prajñā* and his *ṛtambharā prajñā* is based on Buddhism's *bhāvanāmayī prajñā*. The main difference is that Patañjali here adds the dimension of three *pramāṇas*: *pratyakṣa*, *anumāna* and *āgama* (*śruta*). Hence, *śrutamayī prajñā* corresponds to *āgamapramāṇa* (verbal testimony), *cintāmayī prajñā* corresponds to *anumānapramāṇa* (inference) and *bhāvanāmayī prajñā* corresponds to *pratyakṣa pramāṇa* (perception). This perception is supposed to be not ordinary perception but rather extraordinary because it is based on meditation. Hence, it is supposed to give insight into deeper or subtler truths.[110]

The epistemological position suggested by this aphorism and explained in the commentary by Vyāsa is that both inference and words have only universals as their objects, whereas perception has particulars as its object.[111] This position is later found in the works of Diṅnāga and Dharmakīrti. It is not vividly found in the works of Asaṅga and Vasubandhu. However, Vyāsa's position is different from that of Diṅnāga and Dharmakīrti and somewhat continuous with Vasubandhu, insofar as the nature of object is concerned. Vyāsa, while commenting upon YS I.7, says that an object (*artha*) is both universal and particular by nature (*sāmānyaviśeṣātmaka*). This is similar to Vasubandhu's position that things have both *svalakṣaṇa* (own nature) and *sāmānyalakṣaṇa* (universal nature).[112]

तज्जः संस्कारोऽन्यसंस्कारप्रतिबन्धी ।।50।।

Tajjaḥ saṃskāro'nyasaṃskārapratibandhī.

The latent impression born therefrom prohibits other latent impressions.

Comments

Patañjali's idea that the *saṁskāras* (impressions) generated by *prajñā* hinder other *saṁskāras* is consistent with the Buddhist doctrine of detachment towards lower planes (*adhobhūmi-vairāgya*) and that of fetters (*saṁyojana*).

Detachment towards lower planes: According to this doctrine, when an aspirant reaches a higher plane in his spiritual development, he develops detachment towards the objects belonging to the lower plane; but at the same time, he can develop attachment towards the objects belonging to the plane he has reached. So, when he gets rid of earlier latent impressions of a lower plane, he generates new impressions of a higher plane.

Fetters: According to the doctrine of fetters, one does not get rid of all fetters even in a high stage of spiritual development. Vasubandhu acknowledges ten fetters (*saṁyojanas*), out of which five belong to a lower region (*avarabhāgīya*).[113] They are (1) dogmatic belief in self (*satkāyadṛṣṭi*); (2) attachment to mere rules and rituals (*śīlavrataparāmarśa*); (3) doubt (*vicikitsā*); (4) excitement for sensuous pleasure (*kāmacchanda*); and (5) hatred (*vyāpāda*). They are called lower region fetters because they belong to the world of sensuality (*kāmadhātu*) (AKB V.43). By following the path of morality and meditation, one crosses these fetters, develops detachment towards the lower world and goes to the higher worlds: first to the form world (*rūpadhātu*), then to the formless world (*arūpadhātu*). When the aspirant reaches a higher sphere, he is free from the lower region fetters, but now he has to face the upper region fetters. There are five upper region fetters: (6) attachment to the form world (*rūparāga*); (7) attachment to the formless world (*arūparāga*); (8) excitement (*auddhatya*); (9) ego (*māna*); and (10) misconception (*avidyā*). They are called upper region fetters because without abandoning them one does not cross the higher world (AKB V.45).

The idea is that the lower-region fetters are abandoned at the level of form-meditation (*rūpadhyāna*) and one enters the form world (*rūpadhātu*). But form-meditation can generate attachment to the objects belonging to the form world (*rūparāga*). This impression (*saṁskāra*) of form-meditation is confirmed by a still higher meditation, namely formless meditation (*arūpadhyāna*). But formless meditation may generate attachment to the formless world (*arūparāga*). Similarly, belonging to the formless world itself may create an excitement, a self-esteem, and one may not abandon a basic form of misconception (*avidyā*) even in this stage. When these higher region fetters are abandoned, one becomes liberated in true sense.

Patañjali seems to reflect these ideas when he says that there should be cessation of the *saṁskāras* generated by the truth-bearing wisdom (*ṛtambharā prajñā*) as well in order to attain seedless ameditative absorption.

तस्यापि निरोधे सर्वनिरोधान्निर्बीजः समाधिः ।।51।।

Tasyāpi nirodhe sarvanirodhān nirbījaḥ samādhiḥ.

When that (=the latent impression born from the truth-bearing wisdom) too ceases then all latent impressions cease; and the meditative absorption becomes seedless.

Comments

Here Patañjali calls the highest meditative state *nirbīja-samādhi* (seedless trance). This idea also has reference to Buddhism.

As stated in the comments on YS I.46, the term *bīja* used by Patañjali seems to be derived from early Yogācāra Buddhism, where it stands for *ālayavijñāna* (store consciousness which consists of latent impressions (*vāsanās*)).

Nirbīja-samādhi in this sense is the state of meditative absorption in which there are no latent impressions.

It is supposed to be the state of the cessation of all mental states which Buddhist Abhidharma would refer to as the highest state of meditative trance (*saṁjñā-vedayita-nirodha*: cessation of perception as well as sensation), which we discussed while commenting upon YS I.2. Patañjali is suggesting here that in that state, latent impressions also cease. Now, will this state be the state of "embodied liberation" or "disembodied liberation"? Will such a person who attains a seedless meditative trance live as a person? What kind of life will such a person live? It remains a mystery.

Notes

1 *Kaṭhopaniṣad* (6.10–11) defines *yoga* on these lines:

> *yadā pañcāvatiṣṭhante jñānāni manasā saha/Buddhiś ca na viceṣṭate tām āhuḥ paramāṁ gatim//Tāṁ* ***yogam*** *iti manyante sthirām indriyadhāraṇām/Apramattas tadā bhavati yogo hi prabhavāpyayau//*

Here *yoga* is identified with *indriyadhāraṇā*.

2 For a detailed exposition of Asaṅga's use of the term yoga, see Appendix I, Part III. (For Vasubandhu's similar use of the term, see AKB VI.10.)

3 Lossen and Garbe maintained this. See Tandon (1998: v).

4 "*anuśāsanaprātihāryeṇa tu hitena iṣṭena phalena yogo bhavati upāyopadeśāt*" (AKB VII.47).

5 Vasubandhu in AKB VII.11d cites the sixteen-fold classification of *citta* (accepted by Vaibhāṣikas) in the context of "the knowledge of another mind" (*paracittajñāna*). Asaṅga

in *Śrāvakabhūmi* (ŚB: 294) gives a similar but twenty-fold classification in the context of "mindfulness of mind". On the "defiled" side, he adds *sarāga* (attached), *sadveṣa* (hateful) and *samoha* (deluded), and on the "wholesome" side he adds *arāga* (unattached), *vigatadveṣa* (free from hate) and *vigatamoha* (free from delusion). But he does not contain in his list *parītta* (mean) and *mahadgata* (lofty), which Vasubandhu includes.

6 "*vikṣiptaṁ yad bahirdhā pañcasu kāmaguṇeṣu anuvisṛtam*" (ŚB: 297). See also AKB VII.11d.

7 *karmajena hi dhātunā vaiṣamyeṇa vyākulam avaśaṁ bhraṣṭasmṛtikaṁ cittaṁ vartata iti kṣiptm ity ucyate*" (AKB IV.58c).

8 "*mahābhūmikatvāc ca samādheḥ sarvacittānām ekāgratāprasaṅgaḥ. na, durbalatvāt samādheḥ*" (AKB VIII.1).

9 "*tasmād acittikā nirodhasamāpattir iti vaibhāṣikāḥ*" (AKB II.44).

10 "*evam cittam apy asmād eva sendriyāt kāyāj jāyate na cittāt*" (AKB II.44).

11 "*yasya acittikā samāpattis tasyaiṣa doṣo, mama tu sacittikā samāpattir iti*" (AKB II.44).

12 Frauwallner (1984: 323) remarks that Vindhyavāsī (the Sāṅkhya philosopher, whom Patañjali and Vyāsa seem to be following) had given up the doctrine of three psychic organs (*antaḥkaraṇa*) and removed all psychical process to the thinking organ (*manas*). Though this may be correct, consistently using the term *citta* for the thinking organ seems to be the mark of the Buddhist influence.

13 Hence, we find Patañjali's Yoga under the dilemma of *ātmavāda* and *anātmavāda*. See my comments on YS IV.25.

14 So, he asks, "*kasmāt punar ete cittacaittanirodhasvabhāve satyāv asaṁjñisamāpattiḥ saṁjñāvedita-nirodha-samāpattiś cocyete*?" (AKB II.44) ("If these stages are of the nature of *nirodha* of *citta* and *caitta*, why are they called *asaṁjñi-samāpatti* and *saṁjñā-vedayita-nirodha samāpatti*?")

15 "*cittaṁ caittāḥ sahāvaśyam*" (AK II.23) (*citta* and *caitta*s are necessarily together).

16 Woods (2003: 8) translates the aphorism as "Yoga is the restriction of the fluctuations of the mind-stuff". Bronkhorst (1993: 71) translates the definition of *yoga* as "*Yoga* is the suppression of the activities of the mind" and associates it with the main stream meditation in Jaina and Hindu scriptures. However, I treat this definition of Yoga as a mark of Buddhist influence.

17 See Monier-Williams' dictionary on *ni+rudh*.

18 See Pali English Dictionary on *nirujjhati*.

19 For instance, AKB II.62, VI.27, VII.50–51.

20 For the analysis of *Upaniṣads* in this regard, see Jayatilleke (2010), Chapter 1.

21 "*atasmiṁs tad iti pratyayo viparyayaḥ*", "*mithyāpratyayo viparyayaḥ*" (PB: 40).

22 "*asyā eva tathatāyā evam aparijñātatvād bālānāṁ tannidāno'ṣṭavidho vikalpaḥ pravarate trivastujanakaḥ*" (BSB: 34).

23 Vyāsa's examples are: "Consciousness is the nature of puruṣa", "Puruṣa is inactive". The first sentence suggests that *puruṣa* and consciousness are two different things related to each other, when in fact they are identical. The second sentence suggests that inactivity, like other characteristics, is a positive characteristic of puruṣa, when in fact there is absence of activity. His examples only suggest that sentences sometimes mislead us if they are taken at their surface value.

24 This *saṁskāra* is called *bhāvanā* in the Vaiśeṣika system. *Smṛti* in Nyāya-Vaiśeṣika is defined as "*saṁskāramātrajanyaṁ jñānam smṛtiḥ*" (*Smṛti* is the cognition which arises due to *saṁskāra* alone).

25 Mindful awareness can be called awareness of what has happened "just now", that is, just one moment back. It is similar to memory in this sense. Similarly, the Buddhist concept of *smṛti* (Pali: *sati*) is similar to the Jaina concept of *samiti* (attentiveness). For my detailed account, see Gokhale (2007).

26 AKB II.24. Dominic Wujastyk (2016) points out that the word *sampramoṣa* has been used in Buddhist literature of the second century in the sense of forgetting. So *asampramoṣa* simply means "not forgetting". His argument is well taken. The only difficulty in translating the word as not forgetting is that defining memory as not forgetting will make the definition a case of *petitio principi*. I prefer to translate *asampramoṣa*

as non-loss (which is the meaning accepted in Edgerton's Buddhist Hybrid Sanskrit Dictionary. It will make the definition non-trivial).

27 "What is cultivation (*bhāvanā*)? It is the repeated practice (*abhyāsa*) of the eight conditions of abandonment (*prahāṇasaṁskāra*)" (ADS: 73) ("*Bhāvanā katamā? aṣṭānāṁ prahāṇasaṁskārāṇāṁ abhyāsaḥ*"). Here the eight conditions of abandonment are desire (*chanda*), exertion (*vyāyāma*), faith (*śraddhā*), calm (*praśrabdhi*), mindfulness (*smṛti*), circumspection (*samprajanya*), volition (*cetanā*) and equanimity (*upekṣā*).

28 "*catvāro bhāvanāheyāḥ tad yathā rāgaḥ pratigho māno' vidyā ca. dṛṣṭasatyasya paścāt mārgābhyāsena prahāṇāt*" (AK V.5a and AKB on it).

29 "*dīrghakālābhyāsaḥ nirantarābhyāsaḥ satkṛtyābhyāsaś ca*" (AKB VII.34).

30 "The practitioner, who sees forms inside, also sees forms outside, whether they are small, beautiful or ugly. He knows those forms by overcoming them, he sees them by overcoming them; this is the first act of mastery" (AKB VIII.35a). Vasubandhu then extends this idea to other types of objects and talks about eight types of mastery. Asaṅga correlates *abhibhvāyatana* with *vaśavartanatā* ("*abhibhāyataneṣu ālambanam abhibhavati vaśavartanatām upādāya*" (ADS: 96)).

31 Vasubandhu calls it *adhobhūmi-vairāgya* (AKB VIII.14ab).

32 The term *samprajñāta* (*samādhi*) used by Patañjali could have been derived from the term *samprajñāna* (the Pali original of which is *sampajañña*, meaning mindful awareness). As will be shown, *samprajñāna* arises in one of the stages of *rūpadhyāna*. Hence the word *samprajñāta-samādhi* can be interpreted as the meditative absorption in which *samprajñāna* arises.

33 For a detailed account of the Buddhist theory of form-meditation, see Appendix I.

34 "*dhyānāntare vitarkaś ca nāsti*" (AK II.31c and AKB on it).

35 See definition of *asmitā* in YS II.6.

36 Tandon understands the stage of *anya* as *sotāpanna* who has experienced *saṁjñā-vedayita-nirodha*. In the *sotāpanna* state some latent impressions also remain (Tandon, 97: fn. 3).

37 "*vairāgyāt prakṛtilayaḥ*" (SK 45).

38 "*rūpadhātau tu triṣu dhyāneṣu yā navabhūmayaḥ tās tisraḥ sukhopapattayaḥ. te hi devā samādhijena ca prītisukhena ca niṣprītikena ca sukhena sukhaṁ viharanto dīrgham adhvānaṁ tiṣṭhanti*" (AKB III.71).

39 "*Śraddhā cetasaḥ samprasādaḥ . . . śraddadānasya vivekārthino vīryam upajāyate, samupajātavīryasya smṛtir upatiṣṭhate. smṛtyupashāne ca cittam anākulaṁ samādhīyate. samāhitacittasya prajñāviveka upāvartate*" (YB I.20). It may be noted here that the definition of *śraddhā* as "*cetasaḥ samprasādaḥ*" reflects verbatim Vasubandhu's definition (AKB II.25). Similarly, Vyāsa's use of the word *smṛtyupasthāna* is also typically Buddhist.

40 "*bhāvanā katamā? yat śraddhendriyeṇa satyeṣu abhisampratyayasamutthānaṁ prayogabhāvanā. vīryendriyeṇa satyeṣu utpannābhisampratyayasya abhisambodhyarthaṁ vyāyāmasamutthāna-prayogabhāvanā. smṛtīndriyeṇa satyeṣu ārabdhavīryasya smṛtisamprayogārthaṁ asampramoṣasamuthānaprayogabhāvanā. samādhīndriyeṇa satyeṣu samprayuktasmṛteḥ cittaikāgratāprayogabhāvanā. prajñendriyeṇa satyeṣu samāhitacittasya pravicayasamutthānaprayogabhāvanā*" (ADS: 74).

41 "*yo mauladhyānabhūmikāni nirvedhabhāgīyāni utpādayati sa tatraiva janmani satyāni avaśyaṁ paśyati. tīvrasaṁvegatvāt*" (AKB V.22).

42 Here Buddhists maintain that a non-returner, due to his spiritual progress in the *kāma*-world, goes to higher worlds (such as *rūpa* and *ārūpya*), but is not reborn in the *kāma*-world (he could be reborn in higher world); he is ultimately liberated from the higher world only.

43 See ŚB: 70, AKB VI.33. Bahulkar (1985) discusses the two aphorisms (YS I.21–22) in detail. He claims that Vyāsa's interpretation of the aphorisms in terms of nine categories of yogins is far-fetched. He also claims that both Patañjali and Vyāsa are under the influence of Buddhism.

44 "*evaṁ hi bodhisattvāḥ praṇidhānaṁ kurvani, aho batāham andhe loke'pariṇāyake buddho loka utpadyeyam anāthānāṁ nātha iti*" (AKB III.96b).

45 "*trīṇi āvaraṇāni uktāni bhagavatā, karmāvaraṇaṁ, kleśāvaraṇaṁ vipākāvaraṇam ca*" (AKB IV.96: Introduction to the verse).

46 Edgerton points out that the words *āsrava*, *āsaya*, *āśaya*, *anuśaya* and *āsava* often get confused or identified with each other in Pali and Hybrid Sanskrit. See entities on *āśaya*, *āsava*, *anuśaya* and *āsrava*.

47 The whole of *Māṇḍūkyopaniṣad* is about the relation between *Oṁ* and *ātman*. The fifth question in *Praśnopaniṣad* is about the importance of repeating *Oṁ*. In fact, *Oṁkāra* as the symbol for *puruṣa/ātman/brahman* is a frequently occurring theme in *Upaniṣads*.

48 See my comments on YS I.23.

49 "*nanu abhidharme daśa kleṣamahābhūmikāḥ paṭhyante . . . cetaso vikṣepaḥ . . . pramādaś ceti*" (AKB II.26, p. 56, lines 9–11).

50 "*samādhir eva kliṣṭo vikṣepaḥ*" (AKB II.26ab).

51 *adhyātmam upādāya antarāyaḥ* (ŚB: 144).

52 "*dṛṣṭa eva dharme viṣamacārī bhavati. yenāsyābhīkṣṇaṁ vāto vā kupyati pittāṁ vā śleṣmaṁ vā viṣūcikā vā kāye santiṣṭhante*" (ŚB: 144).

53 "*styānaṁ katamat? yā kāyagurutā, cittagurutā, kāyākarmaṇyatā cittākarmaṇyatā*" (AKB p. 56). Compare Vyāsa's definition: "*styānam akarmaṇyatā cittasya*" (YB I.30).

54 "*kāyikaṁ styānaṁ caitasikaṁ styānam ity uktam abhidharme*" (AKB II.26ab).

55 "*agantukāmatā mārgavibhramo mārgasaṁśayaḥ /*
Ity antarāyāś cittasya gamane' tas trideśanā //" (AK V.44).

56 "*kleśamahābhūmikā dharmāḥ*" (AK II.26).

57 ŚB: 147. "*śamathāntarāyaḥ prabhinnaḥ*". Here either the reading is corrupt or the editor's conjecture is incorrect. From the context, it is clear that "*prabhinnaḥ*" should be replaced by "*pramādaḥ*".

58 For example, "eats heavily (*bhojanaguruko bhavati*), enjoys sleeping (*śayanasukhaṁ svīkaroti*), enjoys company of lay-persons (*saṁgaṇikārāmo bhavati*)" and so on. ŚB: 145–6.

59 "*mithyādarśana-avirati-pramāda-kaṣāya-yogā bandhahetavaḥ*" (TAS 8.1).

60 See the comments on YS II.30–31

61 "*aviratiś cittasya viṣayasamprayogātmā gardhaḥ*" (YB I.30).

62 "*Anavasthitacittasya, saddharmam avijānataḥ | pāriplavaprasādasya, prajñā na paripūryate ||*" (UV XXXI.28). The parallel Pali verse can be found in *Dhammapada* 3.6. Thanks to Professor Maithrimurthi for bringing this to my notice.

63 AK II.7–8; "*prītirhi saumanasyam*" (AKB on AK II.8a).

64 "*dveṣo viparyayāt*" (AK V.55b). See also AKB on it.

65 Tandon (1998: 35) in this context assimilates *aṅgamejaya* with the Pali term *phandita* (trembling). Tandon refers to a statement in *Saṁyuttanikāya* 54.7.7: "*ānāpānassatisamādhissa bhikkhave bhāvitattā bahulīkatattā neva kāyassa iñjitattaṁ vā hoti phanditattaṁ*" ("Oh Bhikkhus, when one cultivates and repeatedly practices concentration on mindfulness of breathing, the unsteadiness or trembling of his body never arises").

66 "*yasya tu pratyarthaniyataṁ pratyayamātraṁ kṣaṇikaṁ ca cittaṁ tasya sarvam eva cittam ekāgraṁ nāsty eva vikṣiptam*" (YB I.32).

67 The word *upekṣā* has both senses: indifference and equanimity. The question is in which sense Patañjali has used the word.

68 "*apramāṇāni catvāri vyāpādādivipakṣataḥ/*
maitryadveṣo'pi karuṇā muditā sumanaskatā//
upkekṣā'lobha ākāraḥ sukhitā duḥkhitā bata/
modantāmiti sattvāś ca" (AK VIII.29–30c).

69 "*maitrī kāruṇyam ārteṣu śakye prītir upekṣaṇam / prakṛtistheṣu bhūteśu vaidyavṛttiś caturvidhā //*" (CS, Sūtrasthāna 9.26).

70 "*maitrīpramodyakāruṇyamādhyasthyāni ca sattvaguṇādhikakliśyamānāvineyeṣu*" (TAS 7.6).

71 It should be noted that even in this sense *upekṣā* is not as positive as *maitrī* and *karuṇā* are. It involves a kind of neutrality, detachment, passivity.

72 One can ask: does Asaṅga, by formulating *upekṣā* in terms of the "good of all", indicate that *upekṣā* need not be as neutral as it is supposed to be? Similarly, does Patañjali by saying that *upekṣā* is about sinfulness (that is, removal of sinfulness of all), suggest that it is not neutral or passive as it is supposed to be?

73 See my comments on YS III.23

74 "When one has fixed attention on the tip of the nose, one has experience of divine odour. That is the activity towards odour" (YB I.35). Bronkhorst (1984) accepts this interpretation and thinks that in order to justify the special character of this practice, *sthitinibandhanī* in the aphorism should be taken as a *Bahuvrīhi* compound. Bronkhorst's interpretation is problematic. As a *Bahuvrīhi* compound, it would be *sthitinibandhanā*. Moreover, Vyāsa's interpretation need not be accepted when a more literal interpretation of the aphorism makes sense.

75 The term *jyotiṣmatī* resembles the term *arciṣmatī*, which stands for the fourth plane (*bhūmi*) of the Bodhisattva according to the Mahāyāna Buddhist theory of ten *bhūmis*.

76 Edgerton points out (Buddhist Hybrid Sanskrit Dictionary: the entry on *bhūmi*), following Mahāvyutpatti, that *śrāvakāyāna* accepts seven stages (*bhūmis*) of religious development, of which *vītarāgabhūmi* is the sixth.

77 Narada (1988, Chapter 36).

78 In this context, Professor Maithrimurthi made an interesting suggestion to me that possibly "the mind which has vītarāga as the object" does not mean "the mind which has attachment-free person as an object", but rather "the mind which has attachment-free state (of mind) as the object".

79 Asaṅga explains how one should develop this alertness (ŚB: 105–8).

80 For details see Narada (1988: 519–38)

81 Apte (1957)) gives both the meanings of *abhimata*: (1) desired, wished, dear, beloved and (2) approved, accepted, admitted.

82 In YS I, Patañjali uses the word *vā* in this sense seven times (YS I.23, 34–39)

83 See my comments on YS II.32 and YS II.44.

84 See Appendix II.

85 Tandon (1998: 18) points out on the basis of *Visuddhimagga* (v.28) that the meditation on *kasiṇa* (such as earth *kasiṇa*) enables one to reach the stage of mastery with regard to small and boundless objects.

86 "*vimokṣaprāveśikāni abhibhvāyatanāni. abhibhvāyatanaprāveśikāni kṛtsnāyatanāni*" (AKB VIII.36d). Here the term *abhibhvāyatana* (derived from *abhi+bhū*=to dominate, concur, master) means mastery.

87 Frauwallner (1984: 318) remarks that by the time of Vindhyavāsī (the Sāṅkhya philosopher of the first half of the fifth century CE), some reforms in the classical Sāṅkhya took place, one of which was adoption of the atom-doctrine of the Vaiśeṣika.

88 For example, in Asaṅga's works: ŚB: 447–60; ADS: 68–9. In Vasubandhu's works: AKB VIII.1–6

89 For example, "*paṭhamaṁ jhānaṁ samāpannassa nīvaraṇehi cittaṁ vivittaṁ hoti*" (quoted by Shukla, Karuṇesh in ŚB: 447, footnote 1) (The mind of the one who has attained the first meditation becomes separated from hindrances).

90 "*samāpattiḥ śubhaikāgryam*" (AK VIII.1c)

91 "*abhedena kuśalacittaikāgratā dhyānam. samādhisvabhāvatāt*" (AKB VIII.1c).

92 On the question of how the five consciousnesses associated with gross forms as well as subtle forms (*savitarka-savicāra*) are initially described as "without mental construction" ("*yadi pañca vijñānakāyāḥ savitarkāḥ savicārāḥ, katham avikalpāḥ ity ucyante* ?" AKB I.33), Vasubandhu clarifies that *vikalpa* is of three kinds: (1) mental construction caused by distracted mind (*abhinirūpaṇā-vikalpa*); (2) mental construction caused by memory (*anusmaraṇa-vikalpa*); and (3) matural mental construction (*svabhāva-vikalpa*). The five consciousnesses are without the first and second type of mental construction, not the third.

93 "*tatra svabhāvavikalpo vitarkaḥ*" (AKB I.33).

94 See the comments on YS I.11.

95 For details, see Appendix I and the comments on YS I.17.

96 "*caturthaṁ dhyānam antyam. Tatra catvāry aṅgāni. aduḥkhāsukhā vedanā, upekṣāpariśuddhiḥ,* ***smṛtipariśuddhiḥ*** *samādhiś ca*" (AKB VII.8cd). For details, see Appendix I.

97 For an argument to this effect, see Appendix I.

98 "*cittaudārikatā vitarkaḥ, cittasūkṣmatā vicāraḥ*" (AKB II.33ab).

99 See my comments on YS I.18.
100 Asaṅga in YAB: 11 associates *bīja* with *ālayavijñāna*: "*ālayavijñānaṁ bījāśrayaḥ*".
101 "*nirbhayatā hi vaiśāradyam*" (AKB VII.32). The Buddha was said to possess four-fold fearlessness (*vaiśāradya*) based on four types of knowledge and power.
102 See Table A1.2 included in Appendix I.
103 "*adhyātmasamprasādo nāma ka eṣa dharmaḥ? Vitarka-vicāra-kṣobha-virahāt praśāntavāhitā santater adhyātmasamprasādaḥ sormikeva hi nadī vitarka-vicāra-koṣobhitā santatir aprasannā vartate*" (AKB VIII.9).
104 "*prajñāpra[prā?]sādam āruhya aśocyaḥ śocato janān/ bhūmiṣṭhān iva śailasthaḥ sarvān prājño'nupaśyati//*" [A wise person, having climbed up the palace of wisdom, looks at the ignorant persons like a person on the mountain looks at people on the ground.] The verse consists of the last two lines of the Sanskrit rendering of the three-line Pali verse from *Dhammapada* (2.8). The Sanskrit rendering is found in *Udānavarga* as follows: "*pramādam apramādena yadā nudati paṇḍitaḥ/prajñāprāsādam āruhya tv aśokaḥ śokinīṁ prajām/parvatastha iva bhūmisthān dhīro bālān avekṣate//*" (UV IV.4).
105 "*dvitīye dhyāne catvāry aṅgāni. adhyātmasamprasādaḥ, prītiḥ sukhaṁ cittaikāgratā ca*" (AKB VIII.7cd).
106 "*tṛtīye pañca tūpekṣā smṛtiḥ prajñā sukhaṁ sthitiḥ*" (AK VIII.8ab).
107 Ibid. The word used for wisdom/insight there is *prajñā*. In AKB VIII.8ab, *prajñā* is interpreted as *samprajñāna*.
108 Tandon (1998: 9). According to Tandon, the word "*ṛta*" in "*ṛtambharā prajñā*" refers to "law of nature", which Buddha expressed as "*dhammaniyāmatā*".
109 As Vasubandhu says: "*prajñā śrutādimayī*" (AK VI.15) and explains "*śrutamayī, cintāmayī bhāvanāmayī ca*" (AKB VI.15).
110 Vasubandhu brings out the difference between the three *prajṇā*s in a different way, *śrutamayī* is about a word, *cintāmayī* is about both (word and object) and *bhāvanāmayī* is about the object. This is similar to Patañjali's idea that *ṛtambharā prajñā* is different because it has *viśeṣa* (a particular thing) as its object.
111 Vyāsa's epistemological position in the commentary on YS I.7 is somewhat mitigated because he says that perception *mainly* determines particular aspect and inference *mainly* determines universal character. "*viśeṣāvadhāraṇa-**pradhānā** vṛttiḥ pratyakṣam pramāṇam, . . . sāmānyāvadhāraṇa-**pradhānā** vṛttir anumānam*". (YB I.7)
112 About the Vipassanā meditator, Vasubandhu says, "*kāyaṁ svasāmānyalakṣaṇābhyāṁ parīkṣate, vedanāṁ, cittaṁ, dharmāṁś ca*" (AKB VI.14cd) (He examines body, similarly he examines sensation, mind and phenomena, in terms of their individual characteristics and general characteristics").
113 "*pañcadhāvarabhāgīyam*" (AK V.43a); "*pañcadhaivordhvabhāgīyam*" (AK V.45a).

2

ON MEANS (*SĀDHANAPĀDAḤ*)

In the fifty-five aphorisms of the second chapter, *Sādhanapādaḥ*, Patañjali deals with the means to the final goal (which could be understood at an empirical psychological level as seedless meditative absorption, or at a metaphysical level as isolation (*kaivalya*)). He presents it as the eight-limbed Yoga and also as the Yoga of action (*kriyāyoga*) and places it in the framework of the four foundational concepts, comparable with the four noble truths of Buddhism. He also correlates them with the doctrine of defilements. Here is an aphorism-wise summary of the *Yogasūtra*'s second chapter.

(1–2): Patañjali begins the chapter with the concept of the Yoga of action (*kriyāyoga*) and its advantages, namely achievement of meditative absorption and attenuation of defilement. **(3–14):** He explains the five defilements by regarding *avidyā* as the breeding ground for the remaining four. Then he explains how defilements can lead to latent impressions of actions, thus leading to their fruition in different forms. He also talks about the ways of abandoning defilements. **(15–27):** Patañjali introduces the all-pervasive problem of suffering and the four foundational concepts related to suffering: what is to be abandoned (namely, the suffering yet to come); the cause of what is to be abandoned; its abandonment; and the means to the abandonment. He also states an advantage of the application of means: namely, the seven-fold ultimate wisdom. **(28–29):** Patañjali then introduces the eight limbs of Yoga, the practice of which leads to the removal of impurities; through this, there is illumination of knowledge and finally discriminating knowledge. The eight limbs are: restraints (*yama*), observances (*niyama*), posture (*āsana*), breath-regulation (*prāṇāyāma*), withdrawal of the senses (*pratyāhāra*), fixed attention (*dhāraṇā*), meditation (*dhyāna*) and meditative absorption (*samādhi*). **(30–45):** Patañjali then states the five restraints, five observances, their importance, the method of establishing oneself in them and the advantages of getting established in them. **(46–55):** Patañjali concludes the chapter by introducing three more limbs—posture, breath-regulation and withdrawal of senses—and stating the advantages of getting established in them.

The comments on these aphorisms will reveal how—along with the Sāṅkhya framework and, occasionally, the Jaina theory of right conduct—different concepts and doctrines of Buddhist psychology and soteriology must have influenced Patañjali and Vyāsa. The comments will also reveal how the Buddhist concepts and doctrines are relevant for understanding various topics introduced and explained by Patañjali and Vyāsa.

तपःस्वाध्यायेश्वरप्रणिधानानि क्रियायोगः ।।1।।

Tapaḥsvādhyāyeśvarapraṇidhānāni kriyāyogaḥ.

Austerity, self-study and the resolve regarding the supreme being constitute the yoga of action.

Comments

Here Patañjali is introducing the notion of *kriyāyoga*, or "yoga of action", which can be called a subsection of the eight-fold Yoga. The three devices mentioned here, namely austerity, self-study and the resolve regarding *īśvara*, constitute the last three of five observances (*niyamas*) mentioned in YS II.32. They will be discussed there in detail.

The question remains as to why Patañjali introduces the three principles separately, in advance, when he could instead noted after stating the observances (in YS II.32) that the last three are called *kriyāyoga*. Two reasons seem possible:

(1) Patañjali wants to attach special importance to these three principles, such that one is supposed to start with the practice of these even before he takes up the eight-limbed course on Yoga. One should start tolerating extreme opposites (*tapas*); one should start studying emancipating texts (*svādhyāya*); and one should resolve to attain the highest state of consciousness of the supreme being (*īśvarapraṇidhāna*).

(2) The three principles are important for Patañjali because their practice makes meditative absorption easy and attenuates defilements, as he states in the next aphorism (YS II.2). This provides an occasion for Patañjali to explain his theory of defilements. His theory of defilements (YS II.3–14) leads to the theory of four basic truths related to suffering and emancipation (*heya-heyahetu-hāna-hānopāya*, YS II.15–26) All this forms the basis of the eight-limbed Yoga. It will be seen in due course how both these theories presuppose the Buddhist background.

समाधिभावनार्थः क्लेशतनूकरणार्थश्च ।।2।।

Samādhibhāvanārthaḥ kleśatanūkaraṇārthaś ca.

The yoga of action should be practiced for cultivation of concentration (*samādhi*) and for attenuating the defilements.

Comments

Here Patañjali is stating the purpose served by the yoga of action. According to him, it leads to cultivation of meditative absorption and reduction of defilements.

In Buddhism, the term *samādhi* means one-pointedness of mind;[1] *bhāvanā* means meditative practice, and *kleśa* means defilement. All these meanings are relevant here. The term *tanūkaraṇa* (of *kleśas*) also has a Buddhist background. *Tanu* means thin, small, tiny. *Tanūkaraṇa* means making something small, attenuating or reducing it. Vasubandhu talks about attenuation or reduction (*tanutva*) of the unwholesome roots, namely desire, aversion and delusion (*rāga-dveṣa-moha*), in the context of the stage of "once returner" (*sakṛdāgāmin*).[2]

अविद्यास्मितारागद्वेषाभिनिवेशाः पञ्च क्लेशाः ।।3।।

Avidyāsmitārāgadveṣābhiniveśāḥ pañca kleśāḥ.

The defilements are five: misconception, egoism, attachment, aversion and obsessive clinging (to a view).

Comments

Patañjali is introducing here the doctrine of *kleśas*. In classical Sanskrit, the word *kleśa* is used in the sense of pain.[3] Sometimes it is also taken to mean affliction, that is, something which afflicts or perturbs the mind. ("*kliśnantīti kleśāḥ*").[4] The Sanskrit word *kleśa* (as its Pali analogue *kilesa*) has another (though related) sense in which the word is generally used in Buddhist literature. It is defilement, something which makes the mind polluted, impure. Patañjali is using the term in that sense. Hence, I translate the term *kleśa* as defilement throughout this work.[5]

In *Abhidharmasamuccaya*, Asaṅga identifies *kleśa* with the second noble truth, namely *samudaya* (the cause of suffering).[6] It is relevant to note the definition and classification of *kleśas*. The defining feature of *kleśa*, according to Asaṅga, is lack of peace (*apraśama*). Asaṅga says,

> What is the defining feature (of *kleśa*)? It is the characteristic (of mind) which, when arises, is characterized by non-peace. When it arises, the body-mind series tends to be without peace.[7]

The two meanings of the word *kleśa*, namely "defilement" and "affliction", can be connected in this way: *kleśa* stands for defilement of mind, which afflicts the mind by taking away its peace.

Asaṅga gives a six-fold as well as a ten-fold classification of *kleśa*. The six *kleśas* are: desire (*rāga*), hatred (*pratigha*), pride (*māna*), misconception (*avidyā*), doubt (*vicikitsā*) and the dogmatic view (*dṛṣṭi*).[8] Asaṅga points out that the dogmatic view is of five kinds, and by including them, the whole classification becomes ten-fold.[9] Vasubandhu accepts the same five-fold and ten-fold classification of *kleśas* under the heading "*anuśaya*" (dormant *kleśas*).[10]

TABLE 2.1 Types of defilements according to Pātañjala-yoga and Buddhism

Defilements according to Patañjali	*Defilements according to Asaṅga and Vasubandhu*
avidyā (misconception)	*avidyā* (misconception)
asmitā (egoism)	*māna* (termed as *asmimāna*, *asmitā* [ego, pride])
rāga (desire)	*rāga* (desire)
dveṣa (hatred)	*pratigha* (hatred)
abhiniveśa (dogmatic view)	*dṛṣṭi* (dogmatic view)
	vicikitsā (doubt)

Sources: YS II.4, ADS: 43, AK and AKB V.1

Notes:

(1) Vasubandhu, while explaining different types of *dṛṣṭi* (dogmatic view), mentions "*akāraṇe kāraṇadṛṣṭiḥ*" ("dogmatically believing that something is a cause when in fact it is not the cause") (AKB V.7) as one of them. He describes the same as "*kāraṇābhiniveśa*" (AK V.8). This shows that he is using the terms *dṛṣṭi* and *abhiniveśa* interchangeably.

(2) Patañjali does not give the status of defilement to doubt (*vicikitsā/saṁśaya*), but includes it in the list of obstacles (YS I.30).

It seems obvious that Patañjali's classification of defilements (*kleśa*s) is inspired by the Asaṅga–Vasubandhu classification of them (*kleśa/anuśaya*). This will be clear from Table 2.1.

In his commentary, Vyāsa tries to interpret the general nature and function of the five defilements in Sāṅkhya terminology. For example, he identifies five *kleśa*s with "*pañca-viparyaya*s" following *Sāṅkhyakārikā*, 47–48, which is far-fetched.[11]

अविद्या क्षेत्रमुत्तरेषां प्रसुप्ततनुविच्छिन्नोदाराणाम् ।।4।।

Avidyā kṣetram uttareṣāṃ prasuptatanuvicchinnodārāṇām.

Misconception is the breeding ground for the other defilements whether they are dormant, small, explicit or gross.

Comments

Here Patañjali is explaining the relationship between *avidyā* and other defilements. *Avidyā* is like the field (breeding ground) in which other defilements grow like sprouts. The sprout is dormant (*prasupta*) when it is in the seed form. The seed is first small (*tanu*), then it breaks out (*vicchinna*)[12] and then becomes gross or large (*udāra*) when it grows. Similarly, defilements are dormant, small, explicit (divided, opened), and then they grow up. All this becomes possible because they are grounded in misconception (*avidyā*).

In his commentary, Vyāsa uses this metaphor, but some of his explanations are problematic. He explains *tanu* as similar to a burnt seed (*dagdhabījakalpa*) from which the sprout simply does not grow. He uses the metaphor of "burnt seed" while explaining *tanūkaraṇa* (attenuation) in the aphorism YS II.2 as well. Vyāsa's interpretation seems unsatisfactory, as the words *tanu* and *tanūkaraṇa* do not have that sense.

That *avidyā* is the root cause of other defilements is clear in Buddhist literature from the way the causal chain leading to suffering is discussed. *Avidyā* is the first link in the chain, which in its further causal course generates *tṛṣṇā* (craving), *upādāna* (clinging) and *bhava* (desire to continue one's existence),[13] which cover other defilements.

Asaṅga in *Yogācārabhūmi* emphasizes the close connection between perversion (*viparyāsa*, which is the same as *avidyā*: see the next aphorism) and defilements (*kleśa*). As he says, "defilements are to be understood in three forms: (1) There is a defilement which has perversion as its root (2) There is a defilement, which is the same as perversion and (3) There is a defilement flowing from perversion".[14]

Vyāsa's explanation of the terms *prasupta* (dormant) and *prabodha* (awakening) clearly reflects Vasubandhu's explanation of these terms in AKB.[15]

अनित्याशुचिदुःखानात्मसु नित्यशुचिसुखात्मख्यातिरविद्या ।।5।।
Ānityāśuciduḥkhānātmasu nityaśucisukhātmakhyātir avidyā
Misconception consists in apprehending non-eternal, impure, painful and non-self as eternal, pure, pleasant and Self (respectively).

Comments

Here Patañjali defines *avidyā* as "(wrongly) conceiving of impermanent as permanent, painful as pleasant, non-self as self and impure as pure". This is exactly the Buddhist conception of *viparyāsa*.[16]

It is important to note here that there is a strong conceptual relation between *viparyāsa and avidyā*. As Asaṅga says, "*avidyā* is the root of *viparyāsa*".[17] Here Patañjali seems to go one step further and identify the two.

दृग्दर्शनशक्त्त्योरेकात्मतेवास्मिता ।।6।।
Dṛgdarśanaśaktyor ekātmatevāsmitā.
Egoism is the seeming identity between the power by which one sees (*dṛkśakti*) and the power of seeing (*darśanaśakti*).

Comments

Patañjali's notion of *asmitā* seems to be inspired by the Buddhist concept of *asmimāna*,[18] which means egoistic feeling of the form "I am". In Buddhism, *asmimāna* refers to the false notion of *ātman*. Asaṅga explains, "What is *asmimāna*? It is the elevated state of mind with a dogma of the form 'I and Mine' with respect to the five aggregates of attachment".[19]

Patañjali seems to have accepted the term *asmitā* in the sense of "false notion of 'I'", but gives it a Sāṅkhya turn. According to the Sāṅkhya tradition, the false notion of "I" is caused by misidentification between the categories, namely *puruṣa* and *buddhi*. Patañjali terms the two categories *dṛkśakti* and *darśanaśakti* respectively.

Vyāsa here quotes the view of Pañcaśikha, which says:

> One would conceive of *ātman* ("I" or self) in place of the cognitive faculty (*buddhi*), out of delusion, when one is unable to see *puruṣa*, which is different from the cognitive faculty in terms of its form, character and knowledge.[20]

सुखानुशयी रागः ।।7।।

Sukhānuśayī rāgaḥ.

Desire is that which grows in accord with pleasure.

Comments

Here Patañjali is defining the third defilement, *rāga* (desire). The definition of *rāga* literally follows the statement from the Buddhist *sūtras* quoted by Vasubandhu (AKB II.3), namely "*sukhāyāṁ vedanāyāṁ rāgo'nuśete*" ("Desire grows or becomes intensified in accord with a pleasant sensation").[21] Here the expression "*anuśete*" is the third person present form of the verb "*anu+śī*", which in this context should be taken to mean "to grow in accord with something". The same idea is conveyed by Patañjali by using "*anuśayin*" as the adjective in the sense of "what grows in accord".[22]

In YS(V), Śaṁkara, the commentator, accepts a different version of *Yogasūtra* II.7–8, according to which the aphorism (YS (V) II.7) reads as "*sukhānujanmā rāgaḥ*" ("Desire is that which arises after pleasure"), and the next aphorism (YS (V) II.8) is read as "*duḥkhnujanmā dveṣaḥ*" ("Hatred is that which arises after pain"). Though modern scholars have a tendency to regard the version of *Yogasūtra* accepted by Śaṁkara as the most authentic version, it does not follow that Śaṅkara's versions of these two aphorisms must be the authentic versions and the popular versions the later interpolations. That is because Śaṁkara himself acknowledges the alternative reading of the two aphorisms and does not question its authenticity.[23] On the contrary, he tries to give a plausible interpretation of the popular versions. He interprets the term *sukhānuśayī* as "*sukham anubhavituṁ śīlaṁ yasya sa sukhānuśayī*" ("*Sukhānuśayī* is that which has tendency to experience pleasure"). Through this, he seems to give an etymological explanation of the word *sukhānuśayī* by relating the word *śayī* to *śīla* through having the common root √*śī*. It seems that the connection of the expression *anuśayī* with the verb *anu+śī* or the word *anuśaya* of the Buddhist Sanskrit was not available to Śaṁkara or that searching for such a connection was completely irrelevant for him. It might have been relevant for Vyāsa, as we find Vyāsa using many Buddhist references (though without acknowledging them). But Vyāsa too is silent on the connection of Patañjali's definitions, namely *sukhānuśayī* and *duḥkhānuśayī* with the term *anuśaya* or the verb *anu+śī*.

It seems obvious, however, that the import of the definitions, popularly attributed to Patañjali, cannot be correctly understood without reference to the term *anuśaya* or the verb *anu+śī* of the Buddhist Sanskrit, though the commentators of *Yogasūtra* have not knowingly or unknowingly taken it into account.

दुःखानुशयी द्वेषः ।।8।।

Duḥkhānuśayī dveṣaḥ.

Aversion is that which grows in accord with pain.

Comments

Just as attachment grows in accord with a pleasant sensation, aversion (*dveṣa*, *pratigha*) grows in accord with a painful sensation. The same context from AKB is applicable here.[24]

In Buddhist literature, the word *pratigha* is used in the list of *anuśayas/kleśas*, whereas the word *dveṣa* is used for one of the unwholesome roots (*akuśalamūla*). Vyāsa in his commentary gives *pratigha* as a synonym of *dveṣa*.[25]

As in the case of the previous aphorism, Śaṅkara, the author of Vivaraṇa, accepts an alternative reading of this aphorism: "*duḥkhānujanmā dveṣaḥ*". The comments on the previous aphorism are all relevant here and hence I am not repeating them.

स्वरसवाही विदुषोऽपि तथारूढोऽभिनिवेशः ।।9।।

Svarasavāhī viduṣo'pi tathārūḍho'bhiniveśaḥ.

The dogmatic view is that which, as it flows naturally, arises even in a learned person.

Comments

As we saw under YS II.3, Vasubandhu understands *dṛṣṭi*, that is, a dogmatic view as *abhiniveśa*. Edgerton in the Buddhist Hybrid Sanskrit Dictionary gives the meanings of the term *abhiniveśa* as false belief, erroneous opinion and adherence to something bad. So, when Patañjali says that *abhiniveśa* can naturally arise even in a learned person, it is perfectly intelligible to interpret the term as dogmatic or false belief. The meanings given in Monier-Williams' Dictionary are: application, intentness, study, affection, devotion, determination, tenacity and adherence. "Adherence to something" is common to both sets of meanings.

In his commentary, however, Vyāsa interprets the term as "the hope (*āśīḥ*) that one will live or the fear of death (*maraṇatrāsa*)", which does not fit in the range of meanings. What makes Vyāsa leave the natural course of interpretation and accept a far-fetched meaning could be a matter of investigation. As we have seen, Vasubandhu identifies *dṛṣṭi* (dogmatic view) with *abhiniveśa*. He mentions under the term *dṛṣṭi* five types of dogmatic views: (1) dogmatic belief in "I" and "mine"; (2) belief in the extreme standpoint of annihilation or eternality; (3) to regard something existent as non-existent; (4) to regard something inferior as superior; and (5) to regard non-cause as cause and non-path as path.[26] It also included believing in Sāṅkhya-yoga path as the true path.[27] Vyāsa would not have appreciated it. Vyāsa did not accept that both extremes, namely annihilation and eternality, should be avoided, because the eternalist view was acceptable to him. This might have led Vyāsa to prefer to interpret the term *abhiniveśa* in quite a different way.

It may be argued, in fairness to Vyāsa, that by "the hope for continuation of life or the fear of death", what Vyāsa really meant was "adherence to one's own life" or "adherence to one's own being", and hence it fits in the basic meaning of the term *abhiniveśa*. In that case, Vyāsa's interpretation of *abhiniveśa* could be associated with what is called *bhavatṛṣṇā* (craving for existence); this, according to Buddhism, is supposed to be common to all beings, whether wise or unwise.

ते प्रतिप्रसवहेयाः सूक्ष्माः ||10||

Te pratiprasavaheyāḥ sūkṣmāḥ.

The subtle defilements are to be abandoned by the reverse creation.

Comments

In this aphorism and the next, Patañjali distinguishes between two types of *kleśas*: *pratiprasavaheya* and *dhyānaheya*; that is, those abandoned by "reverse creation" and those abandoned by meditation. This classification of *kleśa*s might have been influenced by Vasubandhu's classification into *darśanaheya-* (or *dṛgheya*)*-anuśaya*s and *bhāvanāheya anuśaya*s.[28] In the Buddhist scheme, *darśanaheya kleśa*s are those defilements which are abandoned by realization of truths (that is, realization of the four noble truths), and *bhāvanāheya kleśa*s are those defilements which are abandoned through meditation. Patañjali seems to have reconstructed the Buddhist division by using the Sāṅkhya category of *pratiprasava* (that is, *pratisarga*—reverse creation). *Pratiprasava* (reverse creation) means the merger of effects into their causes. This process, according to Sāṅkhya, culminates in the final separation of *prakṛti* and *puruṣa*, which stands for liberation (*kaivalya*). As explained in *Sāṅkhyakārikā* (63–8), *buddhi* has eight forms; of these, seven forms cause bondage and only one (namely discriminating knowledge) liberates. When the discriminating knowledge arises, *buddhi* stops producing manifestations, refrains from all other forms and ultimately merges into its original nature. Hence, it seems that in the previous aphorism what Patañjali basically means by *pratiprasava* is the discriminating knowledge through which the "reverse process" takes place.

According to Buddhism, the liberating knowledge is the knowledge of the four noble truths. There are some defilements which are abandoned through the knowledge of the four noble truths. At a higher level, this knowledge includes the realization that all intoxicants are destroyed (*āsavakṣayajñāna*), which implies the realization that subtle defilements, or root-defilements, are also destroyed.

Patañjali here is replacing the Buddhist idea of liberating knowledge with the Sāṅkhya idea of "reverse creation".

ध्यानहेयास्तद्वृत्तयः ।।11।।

Dhyānaheyās tadvṛttayaḥ.

Their manifestations should be abandoned by meditation.

Comments

As stated in the comment on the preceding aphorism, Patañjali's classification of *kleśas* into *pratiprasavaheya* and *dhyānaheya* seems to be inspired by Vasubandhu's distinction between *dṛgheya-anuśayas* (defilements to be abandoned through realization) and *bhāvanāheya anuśayas* (defilements to be abandoned through meditative cultivation). Hence what Patañjali calls *dhyānaheya* seems to be a reflection of what Vasubandhu calls *bhāvanāheya.*

Patañjali's statement that gross forms or manifestations of defilements (*tadvṛttayaḥ*) are to be abandoned by meditation and subtle forms of defilements are to be abandoned by reverse creation seems to be a simplified version of the Buddhist account, which is much more complex.[29]

क्लेशमूलः कर्माशयो दृष्टादृष्टजन्मवेदनीयः ।।12।।

Kleshamūlaḥ karmāśayo dṛṣṭādṛṣṭajanmavedanīyḥ.

The latent impression of actions has its origin in defilements and (its fruition) is experienced in either seen (=present) or unseen (=future) life.

Comments

Patañjali's classification of *karmāśaya* (that is, the latent impressions produced by actions) into *dṛṣṭa-janma-vedanīya* (to be experienced in the present life) and *adṛṣṭa-janma-vedanīya* (to be experienced in the unseen, that is, future life) is based on the four-fold classification of karma given by Buddhism philosophers like Vasubandhu. Vasubandhu's classification is given in Figure 2.1.

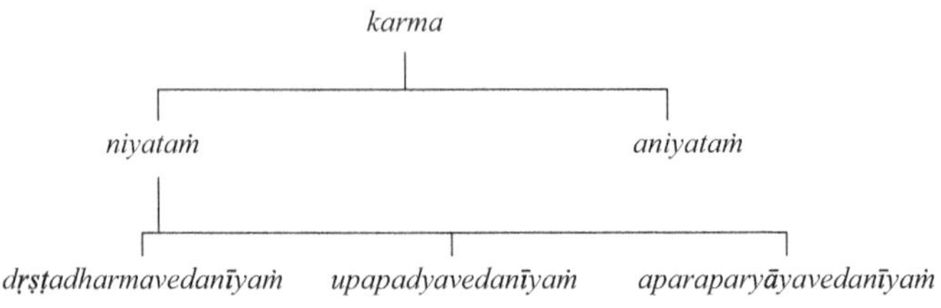

FIGURE 2.1 Vasubandhu's classification of karma

Source: AKB IV.50

Niyataṁ karma: The kind of action, maturation of which will be definitely experienced.
Aniyataṁ karma: It is not certain whether the maturation of this type of action will be experienced.[30]
Dṛṣṭadharmavedanīyaṁ karma: The action which gets matured in the very birth in which it is performed.[31]
Upapadyavedanīyaṁ karma: The action which gets matured in the next birth.[32]
Aparaparyāyavedanīyaṁ karma: The action which gets matured after the next birth.[33]

Patañjali terms *dṛṣṭa-dharma-vedanīyam* as *dṛṣṭajanmavedanīyam*. He does not take into account the *aniyatam* type[34] and combines the two types, namely *upapadyavedanīam* and *aparaparyāyavedanīyam*, into one, that is, into *adṛṣṭajanmavedanīyam*.

Some other details given by Vyāsa in his commentary seem to reflect Vasubandhu's account of the same.

Hence compare:

(1) **Vyāsa:** Hellish beings do not have impressions of actions matured in the same birth.[35]
Vasubandhu: In hell the good karma can have maturation of actions of other three kinds, not the maturation of actions performed in the same life; because there is no "good maturation in the hell".[36]

(2) **Vyāsa:** There is no maturation of action in another birth of those whose defilements are destroyed.[37]
Vasubandhu: Because a noble person is attachment-free person and not subject to degradation, does not perform action, which gets matured in the next or other future births on the same plane.[38]

सति मूले तद्विपाको जात्यायुर्भोगाः ।।13।।

Sati mūle tadvipāko jātyāyurbhogāḥ.

As long as the root (defilement) subsists, its fruition consists in (the kind of) birth, lifespan and (pleasant and painful) experiences (*bhoga*).

Comments

The general rule about fruition/maturation (*vipāka*) of action according to Buddhism is that an action, whether it is good or bad (that is *śubha* or *aśubha*—*kuśala* or *akuśala*), causes maturation, if it contains passions (*āsrava*).[39] The same idea is conveyed by Patañjali by the expressions "*sati mūle tadvipākaḥ*". The classification of the fruits of action as "*jāti*, *āyu* and *bhoga*" reflects the various fruits of action described by Asaṅga in *Yogācārabhūmi*.[40] Similarly, the eight-fold classification of karma found in the Jaina theory of karma is relevant here.[41] Out of the eight types of karma, *gotra-karma* determines "birth in a particular family", *āyuḥkarma* determines the span of life and *vedanīya-karma* determines experiences of pleasure and pain, which can be identified as *bhoga*.

ते ह्लादपरितापफलाः पुण्यापुण्यहेतुत्वात् ।।14।।

Te hlādaparitāpaphalāḥ puṇyāpuṇyahetutvāt.

They result in pleasure or pain as they are caused by merit or demerit (respectively).

Comments

Here Patañjali is suggesting a two-fold classification of karma and its two-fold result, which is central to the doctrine of karma. Accordingly, the conditions such as birth and the span of life, if they are caused by *puṇya* (merit), will result in happiness (or pleasure). If, on the contrary, they are caused by *apuṇya* (demerit), they will result in pain. Here merit and demerit refer to meritorious and de-meritorious actions. This is a simplification of the Buddhist three-fold classification, which is expressed in different ways: (1) *kṣema—akṣema—itarat*, (2) *kuśala—akuśala—itarat* and (3) *puṇya—apuṇya—aniñja*.[42] Hence, Buddhists accept the third category of actions, which can be called neutral (neither good nor bad) and is referred to as *itarat* (also called *avyākṛta*) or *aniñja*. Similarly, though Buddhists would accept that in an ordinary situation a good action (*kuśala-karma*) will lead to pleasure and a bad action (*akuśala-karma*) will lead to pain, they also accept neither pleasure nor pain (neutral sensation) as a possible result of an action. Accordingly, the action, in terms of its maturation, also becomes three-fold: *sukhavedanīya*, *duḥkhavedanīya* and *aduḥkhāsukhavedanīya*.[43] For example, higher meditative states are without pleasure and pain. Hence, according to Buddhists, though generally a good action leads to pleasure, sometimes a "good" action (namely a higher form of meditation) can lead to neither pleasure nor pain.

Patañjali's statement in the earlier aphorism seems to give a simplified or ordinary-level version of the doctrine of action and its results.

परिणामतापसंस्कारदुःखैर्गुणवृत्तिविरोधाच्च दुःखमेव सर्वं विवेकिनः ।।15।।

Pariṇāmatāpasaṅskāraduḥkhair guṇavṛttivirodhāc ca duḥkham eva sarvaṃ vivekinaḥ.

On the account of (the three forms of suffering viz.) that which results (into suffering), that which is pain by nature, and that which is due to conditionedness, and because of the mutual opposition among the modifications of the strands (of *prakṛti*), everything is painful to a discriminating person.

Comments

Here Patañjali is presenting the thesis that all phenomena are *duḥkha* according to a discriminating person. The word *duḥkha* is used here as an adjective and not as a noun. The term *duḥkha* as a noun means pain or suffering. The term *duḥkha* as an adjective means "painful" or "the condition of pain/suffering". (A better translation of the adjectival term, to my mind, is, "unsatisfactory".) Patañjali is here giving two reasons for calling all phenomena as *duḥkha*. The first reason is influenced by Buddhism,[44] the second by Sāṅkhya.

(1) The reason influenced by Buddhism

According to a Buddhist explanation of suffering, every phenomenon is unsatisfactory in one of the three ways:

(1) *Duḥkhaduḥkhatā*: Certain things are unsatisfactory because they are painful in themselves.
(2) *Vipariṇāmaduḥkhatā*: Certain things are unsatisfactory because they result in suffering.
(3) *Saṁskāraduḥkhatā*: Certain things (rather, all conditioned objects) are unsatisfactory because they are conditioned or composite (*saṁskāra*) in nature (and hence are subject to destruction).

This three-fold explanation of unsatisfactoriness is called the doctrine of *trividhaduḥkhatā*.[45] What Patañjali refers to in the earlier aphorism as *pariṇāma(duḥkha)* is *vipariṇāmaduḥkhatā* in Buddhism. What he calls *tāpaduḥkha* is *duḥkhaduḥkhatā* in Buddhism, and what he calls *saṁskāra-duḥkha* is *saṁskāra-duḥkhatā* in Buddhism.

Vyāsa's explanation of the three-fold painfulness is influenced by the discussion in AKB, but it is partly misleading. Particularly, Vasubandhu's explanation of *saṁskāraduḥkhatā* is completely misunderstood by Vyāsa. According to Vasubandhu, *saṁskāraduḥkhatā* means the painful character of all composite things which is due to the very fact that they are composite things.[46] Vasubandhu proceeds to say that only an exalted person can feel this type of painfulness. He then gives the metaphor of hair which is not felt if it is on the palm but creates pain if it is in an eye. An ordinary person is like a palm, whereas an exalted person is very sensitive like an eye.

Vyāsa, while explaining the term *saṁskāraduḥkhatā*, ignores the Buddhist technical sense of the word *saṁskāra* (composite thing or conditioned thing) completely and interprets the term as impression.[47] Secondly, he applies the metaphor of hair, which was restricted by Vasubandhu to *saṁskāraduḥkhatā*, to all types of *duḥkhatā*.[48]

(2) The reason influenced by Sāṅkhya

After giving this Buddhism-based explanation, Patañjali immediately gives a Sāṅkhya-based explanation in terms of conflict between the operations of the strands (*guṇa*) of the *prakṛti*. The idea is that in the liberated state, the three strands are in the state of balance. One experiences suffering when the balance is disturbed. In the state of imbalance, one of the strands dominates the other strands, or two strands come together and dominate the third, and hence there is a conflict among the strands. Since every phenomenon, according to

Sāṅkhya, represents an imbalance among the strands, every phenomenon becomes unsatisfactory.

We can say that this is an abstract metaphysical explanation, based on Sāṅkhya, of the all-pervasive unsatisfactoriness.

Patañjali in this way is placing the empirical (Buddhism-based) and metaphysical (Sāṅkhya-based) explanations side by side, without pointing out any relationship between them.

Vyāsa's commentary of this aphorism also includes the introduction to the next aphorism, which is about the four basic concepts of the system of Yoga. We will take up some of his comments in the discussion of the next aphorism.

हेयं दुःखमनागतम् ।।16।।

Heyaṃ duḥkham anāgatam.

The pain which is yet to come is to be abandoned.

Comments

Through YS II.16 to II.26, Patañjali refers to the four foundational categories of his system. Interestingly, they correspond to the four noble truths found in Buddhism. We know that the Buddha presented his philosophy of life by using the conceptual framework of four noble truths. Patañjali seems to adopt the Buddhist doctrine by using a different but parallel terminology, which is shown in Table 2.2.

Patañjali's four-fold model appears definitely to be a reformulation of the Buddhist model of the four noble truths. Vyāsa, while commenting upon the four-fold model of the Yoga, compares it with the four-fold model of medical science as he says:

> Just as the science of medicine is a complex of four elements: disease, the cause of disease, the healthy condition and the medical treatment, in the same way this science is also a complex of four elements: transmigration, the cause of transmigration, emancipation and the means to emancipation.[49]

It may be noted here that not only the four-fold model of Yoga, consisting of four basic elements, is derived from Buddhism, but also its comparison with the

TABLE 2.2 Four foundational categories according to Buddhism and Yoga

Buddhist terminology	*Patañjali's terminology*
Duḥkha	*Heya*
Duḥkhasamudaya	*Heyahetu*
Duḥkhanirodha	*Hāna*
Mārga	*Hānopāya*

Sources: AK VI.2; YS II.16, 17, 25, 26

model of medical science goes back to the Buddhist *sūtras*. Vasubandhu, while explaining the specific order in which the four truths[50] are presented, states:

> The teaching of the truths follows (the order of) the direct experience (*abhisamaya*). What is the reason why the truths are directly experienced in this way? At the time of deliberation, the truth of suffering, to which one is attached, that which disturbs, from which one wants to be liberated is examined in the beginning. After that the truth of arising (*samudaya-satya*) (is examined) as to what is its cause. Then the truth of cessation (*nirodhasatya*) (is examined) as to what is its cessation. Then the truth of path (*mārgasatya*) is examined as to what is its path. This is similar to investigating into the cause, cessation and remedy of a disease after seeing the disease. The same metaphor is shown for the Truths, in the Sūtras also. In which Sūtra? "The physician accompanied by the four aspects removes the problem".[51]

There is an important difference between the scheme of the four basic truths according to Buddhism and those according to Patañjali. According to the Buddhist scheme, suffering is not described as worthy of abandonment. But it is regarded as worthy of comprehensive understanding (*parijñeya*). The cause of suffering (*duḥkhasamudaya*), on the other hand, is regarded as worthy of abandonment (*heya*).[52]

Patañjali, in his model of the four basic principles, defines the first principle, *heya* (the thing to be abandoned), as "suffering yet to come". This is consistent with the Buddhist approach to suffering, according to which the problem is not how the present suffering will cease but how there will be non-arising (*anutpāda*) of suffering in the future.[53] However, while explaining the other three categories (that is, in aphorisms YS II.17, 25 and 26), Patañjali follows Sāṅkhya metaphysics. This is consistent with Patañjali's "duel policy" of presenting the empirical and the transcendent side by side without relating the two but giving prominence to the transcendent. (He generally depends on Buddhism for the empirical and on Sāṅkhya for the transcendent.) Hence, after giving an empirical explanation of suffering in terms of defilements based on Buddhism, and also the general causal framework of the four noble truths, Patañjali is now pouring Sāṅkhya metaphysical content in them.

Vyāsa, while explaining Patañjali's four-fold model, brings out another important Buddhist reference, that of the two extreme positions (namely eternalism and annihilationism), which the Buddhists of different sects try to avoid. After explaining the four principles related to abandonment (namely, the abandonable, the cause of the abandonable, abandonment and the means to abandonment), Vyāsa asks,

> Here the nature of the one who abandons cannot be worthy of acceptance or abandonment. In the case of its abandonment, the undesirable consequence of annihilationism (*ucchedavāda*) will result. In the case of its acceptance, the doctrine of cause (*hetuvāda*) will result. If both are denied, then eternalism (*śāśvatavāda*) will result. Therefore, this is the right vision.[54]

Vyāsa's statements seem to have reference to Vasubandhu's discussion of *pudgalavāda* (personalism) and *ātmavāda* (the doctrine of soul) in the ninth chapter of AKB. The discussion in brief is like this.

Vātsīputrīyas, the "personalists" (*pudgalavādin*), claim that though *ātman* is rejected, someone who undergoes transmigration (*saṁsartā*), someone who undergoes bondage and liberation has to be accepted.[55] Personalists call that someone a *pudgala* (person). They regard *pudgala* as neither different from the five aggregates nor identical with them, but rather as inexpressible (*avyākṛta*). Personalism, according to them, avoids both the extremes: eternalism and annihilationism.[56]

Vasubandhu is opposed to this view. According to him, accepting *pudgala* is as good as accepting *ātman*, and the latter is a form of eternalism. Like the personalists, Vasubandhu also refers to the two extremes, namely eternalism and annihilationism. The Buddha said, "Oh Ānanda, the statement, '*ātman* exists' leads to eternality; the statement, '*ātman* does not exist (even in conventional sense)' leads to annihilation".[57]

Between the two extremes, namely "there is no 'I' even in conventional sense" and "there is eternal Self", Vasubandhu accepts the middle position according to which the mind (*citta*) itself transmigrates. It is the cause (*hetu*) of recollection,[58] "recollection causes desire, desire causes thought, thought causes effort, effort causes wind, wind causes action".[59]

Having considered in brief Vasubandhu's discussion of *pudgalavāda*, we can make better sense of the Vyāsa's statement of the three positions: (1) annihilationism, (2) the doctrine of cause and (3) eternalism. Vyāsa, in the previously-mentioned statement, seems to be referring to the three positions mentioned by Vasubandhu. Vyāsa's own preference is for eternalism, which is close to *pudgalavāda* and *ātmavāda*. In place of *pudgala* or *ātman*, which Vasubandhu rejects, what Vyāsa seems to have in mind is the *puruṣa* of Sāṅkhya.

द्रष्टृदृश्ययोः संयोगः हेयहेतुः ॥17॥

Dṛṣṭṛdṛśyayoḥ saṃyogaḥ heyahetuḥ.

The conjunction of the seer with the object of seeing is the cause of what is to be abandoned.

Comment

Here Patañjali is giving a causal explanation of suffering according to Sāṅkhya metaphysics. According to Sāṅkhya, the conjunction of *prakṛti* and *puruṣa* is the cause of the creation of the world.[60] And everything (that is, the whole empirical world) is the object of suffering for a discriminating person, according to Patañjali, as we have seen in YS II.15.

प्रकाशक्रियास्थितिशीलं भूतेन्द्रियात्मकं भोगापवर्गार्थं दृश्यम् ॥18॥

Prakāśakriyāsthitiśīlaṃ bhūtendriyātmakaṃ bhogāpavargārthaṃ dṛśyam.

The object of seeing is of the nature of luminosity, activity and inertia. It consists of elements and organs and its purpose is (that *puruṣa* should have) experience (of pleasure and pain) and emancipation.

Comments

From this aphorism onwards up to aphorism YS II.23, Patañjali is giving the explanation from a Sāṅkhya point of view of the terms *dṛśya*, *draṣṭā* and *saṁyoga* used in the previous aphorism. Here he is explaining the term *dṛśya* (the object of seeing, that is, the knowable world). The knowable world is made of the three strands (*guṇas*) of Prakṛti, namely *sattva*, *rajas* and *tamas*. The nature of the three strands is luminosity, activity and inertia respectively. It consists of elements (that is, gross and subtle elements) and organs (which include internal and external sense organs and motor organs). The knowable world, according to Sāṅkhya, is the manifestation of *prakṛti*. Again, according to Sāṅkhya, *prakṛti* manifests itself in various forms, not for itself but for the sake of *puruṣa*. That is, it does so in order that *puruṣa* experiences pleasure and pain (*bhoga*), and that ultimately the *puruṣa* is emancipated.

विशेषाविशेषलिङ्गमात्रालिङ्गानि गुणपर्वाणि ॥19॥

Viśeṣāviśeṣaliṅgamātrāliṅgāni guṇaparvāṇi.

The specific (=gross), the general (=subtle), mark-only and non-mark are the divisions made of the strands (of *prakṛti*).

Comments

Here Patañjali is referring to different divisions of the three strands of *prakṛti*. They are *viśeṣa*, *aviśeṣa*, *liṅgamātra* and *aliṅga*. The five gross elements are called *viśeṣa*, as they are supposed to be specific or concrete. The five subtle elements (popularly called *tanmātras*) are called *aviśeṣa*, as they are supposed to be general, that is, shared features of the gross elements. The manifestations of *prakṛti* are called marks-only (in the sense of inferential marks), because the existence of *prakṛti* is inferred on the basis of them. But *prakṛti* itself is not a manifestation of anything; hence, it is called non-mark.

द्रष्टा दृशिमात्रः शुद्धोऽपि प्रत्ययानुपश्यः ॥20॥

Draṣṭā dṛśimātraḥ śuddho'pi pratyayānupaśyaḥ.

The seer is pure consciousness. Even though pure, he sees (the objects) as reflected in the cognition (*buddhi*).

Comments

Here Patañjali is explaining the term *draṣṭā* (that is, *draṣṭṛ*) as used in aphorism YS II.17. *Draṣṭā* is the same as *puruṣa*, which, according to Sāṅkhya, is of the nature of pure consciousness, which only witnesses without undergoing any transformation. According to Sāṅkhya, *buddhi*, which is the first evolute of *prakṛti*, undergoes transformation according to the object. *Puruṣa* does not undergo any transformation and yet he is said to be the seer. That is because, though *puruṣa* does not see the world directly, he sees it insofar as it is reflected in *buddhi*.

तदर्थ एव दृश्यस्याऽऽत्मा ।।21।।

Tadarth eva dṛśyasyātmā.

For his sake only is the being of the object of seeing.

Comment

This simply means that all the manifestations of *prakṛti* are meant for the sake of *puruṣa*.

कृतार्थं प्रति नष्टमप्यनष्टं तदन्यसाधारणत्वात् ।।22।।

Kṛtārthaṃ prati naṣṭam apy anaṣṭaṃ tadanyasādhāraṇatvāt.

Although the object of seeing is destroyed in relation to him whose purpose has been accomplished, it is not destroyed (completely) because it is common to others.

Comments

Here Patañjali is trying to answer a question about liberation, which can arise in the Sāṅkhya framework. According to Sāṅkhya there are many *puruṣas* and only one *prakṛti*. And as we saw (in YS II.15), the whole world, since it involves conflicting relation among the three strands of *prakṛti*, is an object of suffering. Naturally, when any one *puruṣa* is liberated, the whole suffering should cease, which means that the world should cease to exist. This means that if one *puruṣa* is liberated, all *puruṣa*s should be liberated. As an answer to this possible objection, Patañjali says that at the time of liberation, the world ceases only with respect to the *puruṣa* who is liberated. It continues to exist for other *puruṣas*.

स्वस्वामिशक्त्योः स्वरूपोपलब्धिहेतुः संयोगः ।।23।।

Svasvāmiśaktyoḥ svarūpopalabdhihetuḥ saṁyogaḥ.

The conjunction is caused by the realization of their own nature by them as the powers of the owned (that is, *prakṛti*) and the owner (that is, *puruṣa*).

Comments

The explanation of aphorism YS II.17, which states that the conjunction of the seer with the object of seeing is the cause of what is to be abandoned, comes to the conclusion here. Here Patañjali is explaining the notion of conjunction (*saṁyoga*). What is the conjunction of *prakṛti* and *puruṣa*? Patañjali is saying that it is the result of the (false) identification of them as "owned" (*sva*) and "owner" (*svāmin*). As a matter of fact, there is no such relation between them. But a false idea of the relation of "one belonging to the other" is created. And that is called the conjunction of the two.

तस्य हेतुरविद्या ।।24।।

Tasya hetur avidyā.

The cause of this (conjunction) is ignorance.

Comments

Here Patañjali is concerned with the cause of the cause of suffering, that is, the root cause of suffering. Prima facie this answer appears to be a Buddhist answer according to which *avidyā* (ignorance or misconception) is the first link in the twelve-linked chain of dependent origination. But I think that here Patañjali is concerned with the Sāṅkhya type explanation. So, this *avidyā* seems to be different from the *avidyā* which Patañjali introduces as the basic defilement in aphorism YS II.5. The *avidyā* introduced in the present aphorism is what Sāṅkhya calls *aviveka* (non-discrimination between *prakṛti* and *puruṣa*). This is one of the examples of Patañjali's policy of placing side by side or one after the other the Buddhist and Sāṅkhya ideas without correlating them.

तदभावात्संयोगाभावो हानं तद्दृशेः कैवल्यम् ।।25।।

Tadabhāvāt saṅyogābhāvo hānaṃ tad dṛśeḥ kaivalyam.

The absence of the conjunction which is caused by the absence of that (=ignorance) is the abandonment (of what is to be abandoned). That is the isolation of the seer.

Comments

Here Patañjali is defining the third basic truth of his system, namely *hāna* (abandonment of what is to be abandoned), which is the counterpart of the third noble truth of Buddhism, namely *nirodha* (or *nirvāṇa*). In Buddhism, suffering is said to be caused by craving (*tṛṣṇā*) or, to go still further, by ignorance/misconception (*avidyā*), and naturally *nirodha* (cessation of suffering) is said to be the result of the cessation of craving or that of ignorance. Patañjali is saying here that *hāna* is the result of the removal of *avidyā*. So Patañjali uses the same format of thinking as used in the doctrine of noble truths. But Patañjali is defining both *hāna* and *avidyā* in Sāṅkhya metaphysical terms.

विवेकख्यातिरविप्लवा हानोपायः ।।26।।

Vivekakhyātir aviplavā hānopāyaḥ.

The means of abandoning (what is to be abandoned) is unwavering discriminative knowledge.

Comments

Here Patañjali is defining the fourth basic truth of his system, namely *hānopāya* (the means of abandoning what is to be abandoned), which is the counterpart of the fourth noble truth of Buddhism, namely *mārga*. Again, Patañjali is defining the path here in the Sāṅkhya metaphysical framework. Hence Patañjali says here that unwavering discriminating knowledge is the means to the cessation of suffering. This discrimination is naturally between seer and the object of seeing (*draṣṭā* and *dṛśya*, that is, *puruṣa* and *prakṛti*). Now the Buddhist element will be more visible in the next aphorism.

तस्य सप्तधा प्रान्तभूमिः प्रज्ञा ।।27।।

Tasya saptadhā prāntabhūmiḥ prajñā.

He has the seven-fold ultimate wisdom.

Comments

Patañjali talked about truth-bearing wisdom (*ṛtambharā prajñā*) earlier (YS I.48). Here he is talking about seven-fold ultimate *prajñā*.

Patañjali himself does not explain the nature of the seven-fold wisdom. It is quite possible that his notion of seven-fold wisdom is inspired by the theory presented by Vasubandhu about the seven-fold knowledge which the spiritually advanced beings are supposed to attain. In the seventh chapter of AKB, Vasubandhu discusses at length different types of knowledges the spiritual practitioner experiences at different stages of his pursuit. There he says that a person with attachment can have six types of knowledge.[61] But the attachment-free person has seven types of knowledge.[62] Then he proceeds to say that even an attached person who is pursuing the path of meditation (*bhāvanāmārga*) meditates on seven types of knowledge.[63]

Interestingly, Vasubandhu sticks to the number "seven" while describing the types of knowledge the practitioner has in his various stages. Table 2.3 makes it clear.

Three points are worth mentioning here:

(1) Though Vasubandhu is talking here of seven types of *jñāna* (knowledge) and not seven types of *prajñā* (wisdom), this need not be a serious difference. For Vasubandhu there is a conceptual relation between *prajñā* and *jñāna*. As Vasubandhu says, while distinguishing between *kṣānti* (passive acceptance) and *prajñā* (wisdom), that passive acceptance is a cognitive attitude (*dṛṣṭi*), it is not knowledge, whereas wisdom is both cognitive attitude and knowledge.[64]

(2) The concepts of *dharmajñāna*, *anvayajñāna* and *kṣayajñāna* have obvious reference to the four noble truths. (See the footnotes on these terms.)

(3) While discussing seven types of knowledge, Vasubandhu refers to the stage in which one has concurred all the seven planes (*bhūmis*: seven *bhūmis* are

TABLE 2.3 Seven types of knowledge according to Abhidharma Buddhism

Stage	*The seven types of knowledge*
1 The stage prior to attachment-free stage (*yāvan na vītarāgo bhavati*)	(1) *dharmajñāna* (2) *anvayajñāna* (3) *duḥkhajñāna* (4) *samudayajñāna* (5) *nirodhajñāna* (6) *mārgajñāna* (7) *saṁvṛtijñāna* (AKB VII.22cd)
2 Attachment-free person (*vītarāga*)	(1) to (6) as above and (7) *paracittajñāna* (telepathy) (AKB VII.22b)
3 The penetration of unshakable (*akopya*) stage	(1) to (6) as above and (7) *kṣayajñāna* (AKB VII.23)
4 Detachment at the top of the world (*bhavāgravairāgya*)	(1) to (6) as above and (7) *paracittajñāna* (telepathy) (AKB VII.23)

Sources: AK and AKB VII.22–23

Explanations:

(1) *Dharmajñāna* (knowledge of the phenomena) has the following objects: suffering belonging to the world of sensuality, its cause, cessation and the contrary views (*pratipakṣa*) (*dharmajñānasya kāmāvacaraṁ duḥkhaṁ tatsamudayanirodhapratipakṣāś ca ālambanam*) (AKB VII.3c). Here "contrary views" can be understood as the views contrary to the cause of suffering. Hence, they can be understood as a part of the "path". In this way *dharmajñāna* has a clear reference to the four noble truths. (For *pratipakṣa*, also see the comments on YS II.33.)

(2) *Anvayajñāna* (the subsequent knowledge) has the following objects: suffering belonging to the higher worlds (that is, form and formless worlds), its cause, cessation and the contrary views (AKB VII.3cd). This shows that *anvayajñāna*, too, has a clear reference to the four noble truths.

(3) *Saṁvṛtijñāna* (the knowledge of the conventional world) means the knowledge of all (conditioned as well as unconditioned) objects.

(4) *Kṣayajñāna* is the knowledge of the form "I have known suffering fully. I have abandoned the cause of suffering. I have realized the cessation. I have practiced the path" (AK and AKB VII.7).

> the four *dhyānas* and the three formless spheres (*ārūpyas*)).[65] Now the practitioner is in the eighth plane, which is the ultimate plane. In this sense, it can be called *prāntabhūmi*. Hence the wisdom in that stage can be called *prāntabhūmiḥ prajñā*. It is highly probable, therefore, that when Patañjali talks about seven-fold *prāntabhūmi-prajñā*, he intends to refer to this stage in which one crosses all the seven *bhūmis* and processes seven types of knowledge.

Hence, it seems highly probable that Patañjali's concept of "*saptadhā prāntabhūmiḥ prajñā*" is a reflection of Vasubandhu's "seven knowledges which arise in the ultimate plane".

Vyāsa in his explanation of seven-fold wisdom does not refer to "seven knowledges" explained by Vasubandhu, but his explanation too is partly based on the doctrine of four noble truths. Vyāsa divides the seven types of wisdom referred to by Patañjali into two groups: the group of four and the group of three. The group of four as explained by Vyāsa refers to what Vasubandhu calls *kṣayajñāna/anutpādajñāna*.[66] A comparative table (Table 2.4) makes it clear.

Vyāsa calls this four-fold wisdom as *prajñāvimukti*. Then Vyāsa states the group of the remaining three types of wisdom, which he calls *cetovimukti*. The

TABLE 2.4 Four types of knowledge designating emancipation according to Vyāsa and Vasubandhu

	Vyāsa's first four kinds of prajñā	*Vasubandhu's four kinds of* jñāna: kṣayajñāna *and* anutpādajñāna
1	What is to be abandoned is fully known, nothing of it remains to be known.	I have known the suffering fully; it does not remain to be known fully.
2	The causes of what is to be abandoned are destroyed and they do not remain to be destroyed.	The cause of suffering is fully abandoned and it does not remain to be abandoned.
3	The abandonment is realized by the concentration on cessation.	The cessation is realized and nothing of it remains to be realized.
4	The means to abandonment is practiced.	The path to the cessation is practiced and it does not remain to be practiced.

Sources: YB II.27; AK and AKB VII.7

classification of liberation into *prajñāvimukti* (cognitive liberation) and *cetovimukti* (mental liberation) is very much Buddhist. Vasubandhu refers to the statement in Buddhist *sūtra* literature to this effect.[67] The three types of *prajñā*, which Vyāsa states under *cetovimukti* are the three Sāṅkhya-type descriptions of liberated state (what Vyāsa calls the state beyond the strands (*guṇātīta*) are: (5) *Buddhi* has fulfilled its function, (6) *Guṇa*s merge into their causes and (7) in this state *puruṣa* shines in its original isolated position.[68]

योगाङ्गानुष्ठानादशुद्धिक्षये ज्ञानदीप्तिराविवेकख्यातेः ।।28।।

Yogāṅgānuṣṭhānād aśuddhikṣaye jñānadīptir āvivekakhyāteḥ.

By practicing the limbs of Yoga, the impurities are destroyed. Then there arises illumination of knowledge culminating in discriminative knowledge.

Comments

Here Patañjali is referring to the practice of the (eight) limbs of Yoga, which has a background in the Buddhist idea of the practice of the eight limbs of the noble path. A broad point of similarity between the two approaches to the "Yoga/path" is that they understand it to be the path of "purification". In Buddhism it is called the path of purification (*visuddhimagga*) and in Pātañjala-yoga it is called the means to destruction of impurities ("*yogāṅgānuṣṭhānād aśuddhikṣaye*", YS II.28). Patañjali also talks about the ultimate destination of the path, namely the discriminating knowledge. This destination is the metaphysical knowledge according to Sāṅkhya.

It seems that according to Patañjali, since he gives ultimacy to Sāṅkhya metaphysics, "purification of mind" is a penultimate goal and discriminating knowledge (*vivekakhyāti*) is the ultimate goal. Hence, the empirical, psychological path of Buddhism becomes the means to the metaphysical goal of Sāṅkhya.

Here Vyāsa, while explaining the causal connection between the practice of the eight-limbed yoga and discriminative knowledge, introduces a nine-fold classification of causes. This classification seems to be a subset of the twenty-fold classification of causes given by Asaṅga in *Abhidharmasamuccaya* (ADS: 28–29). Tables 2.5 and 2.6 make it clear.

Observations

From Tables 2.5 and 2.6 it seems clear that there is striking similarity between the types of causes in the Vyāsa list and the Asaṅga list. Vyāsa might have short-listed the Asaṅga list by avoiding repetition and omitting certain types typically relevant for the Buddhist theory of dependent origination.

यमनियमासनप्राणायामप्रत्याहारधारणाध्यानसमाधयो ऽष्टावङ्गानि ।।29।।

Yamaniyamāsanaprāṇāyāmapratyāhāradhāraṇādhyānasamādhayo'ṣṭāv aṅgāni.

Restraint, observance, posture, breath-regulation, withdrawal of the senses, fixed attention, meditation and meditative absorption are the eight limbs of Yoga.

Comments

Here Patañjali is stating the eight limbs of the yogic practice or the path of Yoga. The Buddhist path towards the cessation of suffering is also eight-limbed ("*ārya-aṣṭāṅgika-mārga*").

There is no one to one correspondence between the limbs of the two paths. However, there is an essential similarity between the two ways of understanding the path.

The core similarity between the two is that both the paths can be mapped on the model of three-fold training (*tisraḥ śikṣāḥ*):[69] training of the conduct (*adhiśīlaṁ śikṣā*), training of the mind (*adhicittaṁ śikṣā*) and training of understanding/insight (*adhiprajñaṁ śikṣā*).

The eight limbs of the Buddha's path are right (understanding, thoughts, speech, action, livelihood, efforts, mindfulness and concentration). The first two limbs among them can be subsumed under training of understanding/insight, the next three under training of conduct and the last three under training of mind.

In actual practice of spiritual development, the order can change; there regulation of conduct can come first, then that of mind and finally that of insight. Asaṅga accepts this order. He explains the significance of this order as follows:

> What is the sequential order among these trainings? One whose conduct (*śīla*) is very pure, does not (have to remorse). One who does not (have to) remorse, has joy, pleasure, peace, and happiness. A happy person attains mental concentration (*citta-samādhi*). The person with concentrated mind understands things as they are (*yathābhūtaṁ prajānāti*). . . . In this way these rules of conduct, when practiced with priority lead to the final, unattached release. This is the sequential order of the trainings.[70]

TABLE 2.5 Types of causes according to Vyāsa and Asaṅga

Vyāsa types of causes	*Vyāsa examples*	*Asaṅga types of causes*	*Asaṅga examples*
Utpatti (Origination)	Manas causes a cognition	*Utpatti*	Sense, cognition and the other conditions cause a cognition
Sthiti (Stabilizing cause)	Soul-orientedness stabilizes mind; food stabilizes body	*Sthiti*	Food stabilizes living beings and those who want to be born
Abhivyakti (Manifesting cause)	Light manifests the form; similarly, cognition (of form) manifests the form	*Prakāśa* (illuminating cause)	Light illuminates the forms
Vikāra (Transforming cause)	A divergent object changes mind, or fire transforms the object of cooking	*Vikāra*	Fire transforms the fuel
Pratyaya (Cognitive cause)	Cognition of smoke causes the cognition of fire	*Sampratyaya* (cognitive cause)	Smoke causes the cognition of fire
Prāpti (Leading cause)	Practice of the eight-limbed yoga leads to attainment of discriminating knowledge	*Prāpaṇa*	The path leads to attainment of *nirvāṇa*
Viyoga (Dissociating/ dividing/ cutting cause)	The same practice (as above) destroys impurities	*Viyoga*	Knife cuts the thing to be cut
Anyatvakāraṇa Altering cause or differentiating cause	(i) The goldsmith alters (transforms) the gold (ii) The perception of a woman becomes confusing due to ignorance, painful due to aversion, pleasant due to desire and neutral due to philosophical knowledge	*Pratiniyama kāraṇa* (differentiating cause)	Five types of causal conditions cause five types of post-mortal states (*gati*)
Dhṛti Holding cause	Body holds the organs and the organs hold the body; the gross elements hold bodies and each other. Animals, humans and deities act for each other	*Dhṛti* (holding cause)	The earth holds living beings

Sources: ADS 28–29, YB II.28

TABLE 2.6 Additional eleven types in Asaṅga's list

Types of causes	*Examples*	*Remarks*
Pariṇatikāraṇa (Transforming cause)	Artisan's (goldsmith's?) place transforms gold	Can be understood as *anyatvakāraṇa* of Vyāsa's list
Sampratyāyana (Argument as cause)	Declaration, reason and example cause the knowledge of probandum	Extension of *sampratyaya-kāraṇa* (Asaṅga)/*pratyaya-kāraṇa* (Vyāsa) to the case of inference for others
Vyavahārakāraṇa (Conventional cause)	Signs and view point cause conventional/ communicative practice	It has special significance for Buddhism
Apekṣā (Cause as the object of dependence)	Depending on hunger one wants to eat	Additional type: specially significant for the Buddhist doctrine of dependent origination
Ākṣepakāraṇa (Distant cause)	Ignorance is the cause of decay and death	Typical Buddhist, referring to twelve-linked chain; but can be generalized
Abhinirvṛtti (Accomplishing/ immediate cause)	Ignorance causes disposition	Typical Buddhist, referring to twelve-linked chain, but can be generalized
Parigraha (Other related causes)	The field, water and well etc. cause the rise of crops	This cause overlaps with *sahakārikāraṇa*
Āvāhakakāraṇa (Carrying cause)	Serving a king in a proper way brings about king's satisfaction	This is human/ psychological cause rather than material
Sahakārikāraṇa Associating causes (Totality of caused conditions)	Sense organ, object, and concept are the causes of the cognition	This overlaps with many other causes
Virodhakāraṇa Obstacle as cause (of destruction)	The fire of lightening destroys the crops	This overlaps with *vigogakāraṇa*
Avirodhakāraṇa (Absence of obstacle)	Absence of the fire of lightening helps the rise of crops	Later on, it was called *pratibandhakābhāva* in Nyāya

Source: ADS 28–29

Patañjali seems to follow this order in the arrangement of the eight limbs. Hence the first two (restraints (*yama*) and observances (*niyama*)) can be broadly subsumed under right conduct(*śīla*) and the last three (fixed attention (*dhāraṇā*), meditation (*dhyāna*) and meditative absorption (*samādhi*)) under mental training (*adhicittaṁ śikṣā*). The three in the middle (posture (*āsana*), regulation of breath (*prāṇāyāma*) and withdrawal of senses (*pratyāhāra*)) can be understood as the preparations for mental training.

In Patañjali's eight-limbed Yoga, *prajñā*, or its equivalent, is not mentioned directly. But as we have seen in Chapter 1 (YS I.46–51), *ṛtambharā prajñā* is regarded as a result of the constant practice of *sabīja-samādhi* and it is also regarded as the road to *nirbīja-samādhi* (seedless absorption, which is the same as the final release).

अहिंसासत्यास्तेयब्रह्मचर्यापरिग्रहा यमाः ।।30।।

Ahiṁsāsatyāsteyabrahmacharyāparigrahā yamāḥ.

Non-injury, truthfulness, non-stealing, continence and non-possession are the (five) types of restraint.

Comments

Patañjali is giving here the five-fold classification of restraints. The basic idea behind "restraint" (*yama*) is restraining from or abstaining from or avoiding what one ought not do from a moral and spiritual point of view. Such rules of conduct applicable to all humans are found in different Brahmanical texts where they are termed as *sāmānya-dharma*, *sādhāraṇa-dharma* or *sāmāsika-dharma*. The term used for such moral-spiritual rules of conduct in Buddhism is *śīla* (good conduct) and in Jainism, *vrata* (vow).[71] These rules are generally formulated in terms of *viramaṇa* ("refraining from").

The Yoga classification of *yamas* matches accurately with the Jaina classification of vows (*vrata*). But the five-fold classification of *yamas* (Yoga) and *vratas* (Jainism) also overlaps with the Buddhist five-fold classification of *śīla*. Table 2.7 makes it clear.

Vyāsa's explanation of the five restraints broadly follows the Jaina explanation of the corresponding five vows. The similarities between are seen in Table 2.8. (The Jaina explanation is based on the *Tattvārthasūtra* of Umāsvāti.)

There is one more point of similarity between the Vyāsa's approach to the five restraints reflected in YB II.30 and the Jaina approach to the five vows. Both attach utmost importance to *ahiṁsā*.[72]

TABLE 2.7 *Vrata* (Jainism), *Yama* (Yoga) and *Śīla* (Buddhism)

Jainism (vrata)	*Yoga* (yama)	*Buddhism* (śīla)
hiṁsāvirati	*ahiṁsā*	*prāṇātipātaviramaṇa*
anṛtavirati	*satya*	*mṛṣāvādaviramaṇa*
steyavirati	*asteya*	*adattādānaviramaṇa*
abrahmavirati	*brahmacarya*	*kāmeṣu mithyācāra-viramaṇa*
parigrahavirati	*aparigraha*	Xxxx
Xxxx	Xxxx	*pramādasthānaviramaṇa* (avoiding intoxicating things)

Sources: TAS 7.1; YS II.30; Pali English Dictionary ("*Pañca-sīla*") under the entry on "*śīla*"

TABLE 2.8 The nature of *yamas/vratas* according to Vyāsa and Umāsvāti

Restraint/vow	*Vyāsa's explanation*	*Umāsvāti's explanation*
Ahiṁsā	(1) Not injuring any being any time in any way. (2) Refraining from the causes of injury caused by carelessness.	Refraining from injury. Injury means taking away life out of carelessness. (TAS 7.8)
Satya	Truthfulness means veracious speech and mind.	Refraining from untruth (*anṛta*). Untruth means saying what is not real. (TS 7.9)
Asteya	Stealing means accepting money/goods from another person without scriptural sanction. Not accepting them due to non-desire is non-stealing.	Refraining from stealing. Stealing means accepting what is not given. (TAS 7.10)
Brahmacarya	Celibacy means control over the private organ, namely the generative organ.	Refraining from non-celibacy. Non-celibacy means sexual union. (TAS 7.11)
Aparigraha	Non-possession means not accepting things realizing defects in their acquisition, preservation, destruction, attachment and violence caused by them.	Refraining from possession. Possession means attachment (*mūrchā*). (TAS 7.12)

Sources: YB II.30, TAS 7.8–12

जातिदेशकालसमयानवच्छिन्नाः सार्वभौमा महाव्रतम् ॥31॥

Jātideśakālasamayānavacchinnāḥ sārvabhaumā mahā-vratam.

They constitute a great vow as they are supposed to cover all objects being unrestricted by birth (caste), place, time or conditions.

Comments

Here Patañjali is bringing out the universal and unconditional character of the five restraints. It is interesting to note that just as the five *yamas* are identical with the five *vratas* of Jainism, the notion of great vow (*mahāvrata*) associated with it also seems to be derived from Jainism. Jainas distinguish between small vows (*aṇuvrata*) and great vows (*mahāvrata*). The same vows, when practiced partially, are called small vows, but when practiced universally, are called great vows.[73] Small vows are applicable to house-holders (*agārī*)[74] who cannot follow the rules of conduct with full intensity. The great vows are to be followed by mendicants (*anagārī*) who are supposed to follow the rules of conduct with full intensity—in their universal and unconditional form. Patañjali seems to be following this Jaina approach here with reference to Yoga practitioners. He is suggesting here that a yogin has to follow the moral-spiritual rules of conduct in universal and unconditional form like the "great vows" of the Jainas.

शौचसन्तोषतपः स्वाध्यायेश्वरप्रणिधानानि नियमाः ।।32।।

Śaucasantoṣatapaḥsvādhyāyeśvarapraṇidhānāni niyamāḥ.

Cleanliness, contentment, austerities, self-study and resolve regarding the supreme being are the observances.

Comments

Whereas *yamas* are of the nature of abstentions, *niyamas* are observances. They are positive attitudes. According to Patañjali, a yogin should keep himself clean (physically as well as mentally), maintain contentment of mind, practice various austerities, engage in the study of moral and spiritual disciplines and make resolve to become like the supreme being.

Out of these five observances, *śauca, santoṣa* and *svādhyāya* have a clear background in Buddhism. *Tapas* and *īśvarapraṇidhāna* can be correlated with the Buddhist tradition with some qualifications.

***Śauca*:** Vasubandhu mentions Buddhist *sūtras* referring to three-fold cleanliness ("*Trīṇi śauceyāni*"): cleanliness of body, speech and mind. (AK 4.64) Vyāsa distinguishes between external and internal cleanliness. This seems to have the background in the three-fold Buddhist classification.

***Santoṣa*:** Vyāsa defines *santoṣa* as "not desiring to take more than the available means".[75] We find this idea implied in Vasubandhu's discussion of *santuṣṭi* (contentedness) as a quality useful for successful meditation. He contrasts *santuṣṭi* with *asantuṣṭi* and defines the latter as "desiring more than what one gets".[76] According to him, meditation becomes available to one who keeps away (*vyapakarṣa*) physically and mentally from unwholesome attitudes (*akuśalavitarka*). And keeping oneself away from them in both ways (*vyapakarṣa-dvaya*) is easy for one who initially does not desire much and when he gets something, does not desire more than what he gets. In other words, one who is *alpeccha* (having less desires) and is contented (*santuṣṭa*).[77]

***Tapas*:** *Tapas* is generally understood as penance which involves the acts of deliberately experiencing pain. We find that such acts were performed by sages in the Brahmanical tradition and some such acts of austerity are found prescribed in the Jaina tradition. The Buddha, however, while prescribing the middle path, opposed both the extremes: indulgence in sensuous pleasures and self-mortification.

It seems however that *tapas* in the sense of sustained and rigorous efforts which do not involve self-mortification were acceptable to the Buddhists. There are statements in the Buddhist literature that "The *tapas* of the Buddhas was pleasant (and not painful)".[78] "The noble ones should not perform painful austerities".[79]

What Patañjali means exactly by *tapas* when he includes it in the list of observances is not clear from the *Yogasūtra* itself. Vyāsa defines *tapas* as "tolerating opposite pairs" (*dvandvasahanam*). If these opposite pairs are supposed to be imposed on oneself deliberately, then that is not consistent with the spirit of Buddhism. However, tolerating extreme opposites as they come naturally is an important principle in Buddhism. This principle in Buddhism is called *kṣamā* or *kṣānti*. It does not involve self-mortification. It is quite possible that Patañjali's notion of *tapas* is closer to that of Buddhism than to that of Jainism, which involves self-mortification.

Svādhyāya: Patañjali's concept of self-study (*svādhyāya*), too, has a background in Buddhism. Asaṅga, while elaborating on awareness (*samprajñāna*) in each walk of life, recommends rising early from bed and engaging with commentarial texts or the practice of self-study (*svādhyāya*).[80] By *svādhyāya* he means studying Buddha's teachings with devotion. Vyāsa's interpretation of the term *svādhyāya* as "study of the sciences of emancipation" seems to be a generalized form of the Asaṅga concept, which makes the idea of self-study more inclusive and not sect-specific. But Vyāsa in fact gives two alternative interpretations of the term: (1) study of the sciences of emancipation and (2) repetition of the spell "*Om*". The second interpretation does not seem to be correct because it is covered by *īśvarapraṇidhāna* as suggested by the aphorism YS I.28.

Īśvarapraṇidhāna: We have already discussed the Patañjali's conceptions of the supreme being and the resolve regarding the supreme being (*īśvara and īśvarapraṇidhāna*) while commenting on the aphorisms YS I.24–28.

वितर्कबाधने प्रतिपक्षभावनम् ।।33।।

Vitarkabādhane pratipakṣabhāvanam.

When one is vitiated by unwholesome thoughts, cultivation of an antidote[81] is to be made.

Comments

Patañjali here is referring to "cultivation of the antidote" (*pratipakṣabhāvana*) as the technique which is used for obstructing or removing immoral tendencies.[82]

Vitarka: *Vitarka* by itself is a neutral concept. *Vitarka* means thought (or the gross object of thought). This is how the term is ordinarily used in the context of the first *dhyāna* (associated with *vitarka* and *vicāra*).[83] But, here Patañjali uses the word *vitarka* in a different sense. The Buddhist background can be of help in understanding this other sense.

Buddhists distinguish between two kinds of *vitarka*—*kuśalavitarka* (wholesome thought) and *akuśalavitarka* (unwholesome thought). But in a certain context they use the term *vitarka* in the sense of *akuśalavitarka* only. Interestingly, Asaṅga, while explaining the term *vitarka* as one of the veils, in *Śrāvakabhūmi*, defines it as defiled thoughts such as desirous thought (*kāmavitarka*).[84] Patañjali's use of the word *vitarka* in "*vitarkabādhana*" seems to follow Asaṅga here.

The idea that *pratipakṣabhāvanā* is a means to removal of *vitarka* is clearly present in *Śrāvakabhūmi*. While explaining the method of purifying the mind by removing the veils (*āvaraṇaviśuddhi*), Asaṅga mentions four veils (*āvaraṇa*): distress (*paritamanā*), hindrances (*nivaraṇa*), unwholesome thought (*vitarka*) and dogmatic adherence to *ātman* (*ātmasaṁpragraha*). He states (ŚB: 398) that there are four ways of purifying these veils: by knowing the nature of the veils (*svabhāvaparijñāna*), by knowing the causes (*nidāna*) of the veils, by knowing the defects (*ādīnava*) of the veils and by cultivation of the antidote (*pratipakṣabhāvanā*). This implies that the idea that *pratipakṣabhāvanā* is a means to removal (or avoidance) of *vitarka* was clearly present in Asaṅga, and Patañjali might have adopted it from him.

Pratipakṣabhāvanā

In Abhidharma Buddhism, *pratipakṣabhāvanā* is a part of the theory of fourfold *bhāvanā*. As Asaṅga (ADS: 70–71) and Vasubandhu (AK and AKB VII.27) explain it:

There are four types of cultivation (through contemplation): cultivation of obtaining (*pratilambhabhāvanā*), cultivation of maintaining (*niṣevaṇabhāvanā*), cultivation of the antidote (*pratipakṣabhāvanā*) and cultivation of washing out (*vinirdhāvanabhāvanā*). The first two apply to wholesome qualities and the last two to unwholesome qualities.

(1) The wholesome qualities which one does not have are to be obtained (*pratilambhabhāvanā*).
(2) The wholesome qualities which one has are to be maintained (*niṣevaṇabhāvanā*).
(3) About the unwholesome qualities which one does not have, the contemplation of the antidote is to be cultivated (*pratipakṣabhāvanā*).
(4) The unwholesome qualities which one has are to be washed out (*vinirdhāvanabhāvanā*).

According to the previous classification, *pratipakṣabhāvanā* is to be used to prevent the entry to unwholesome qualities. If this is what Patañjali also means, then *vitarkabādhane* can better be interpreted as "with regard to obstruction (or prevention) of the defiled thoughts".

Patañjali's conception of *pratipakṣabhāvana*, in this way, can be understood better against the background of the Buddhist theory of mental purification as explained by Asaṅga and Vasubandhu.[85]

वितर्का हिंसादयः कृतकारितानुमोदिता लोभक्रोधमोहपूर्वका मृदुमध्याधिमात्रा दुःखाज्ञानानन्तफला इति प्रतिपक्षभावनम् ।।34।।

Vitarkā hiṁsādayaḥ kṛtakāritānumoditā lobhakrodhamohapūrvakā mṛdumadhyādhimātrā duḥkhājñānānantaphalā iti pratipakṣabhāvanam.

The cultivation of an antidote may be done by contemplating that the unwholesome thoughts such as those of injury, whether done or caused to be done or approved, are caused by greed, anger and delusion. Whether they are mild, medium or vehement in degree, they result in the endless fruition of suffering and ignorance.

Comments

Here Patañjali is explaining the statement he made in the previous aphorism. He said there that *pratipakṣabhāvana* (contemplation of an antidote) is to be made in order to obstruct or prevent *vitarka* (the defiled thoughts). In this aphorism, he explains the term *vitarka* and describes the method of *pratipakṣabhāvana*. Both these are influenced by the Buddhist conceptions of *vitarka* and *pratipakṣa-bhāvanā*.

As stated in the comments on the preceding aphorism, the word *vitarka* is used here not in the sense of "gross thought" or "gross object of thought" but in the sense of "unwholesome thought" or "defiled thought". Hence Patañjali says "*vitarkā hiṁsādayaḥ*". This is influenced by the Buddhist classification of unwholesome *vitarka* (*akuśalavitarka* or *kliṣṭavitarka*) into "*kāmavitarka*, *vyāpādavitarka* and *vihiṁsāvitarka*" (that is, the thought of sensuous desire, the thought of hatred, the thought of violence).[86] Here it should be noted that when Patañjali says that "*vitarkā hiṁsādayaḥ*" it should not be taken to mean that "violence (*hiṁsā*) etc. are *vitarkas*" but that "the thought of violence etc. are *vitarkas*".

The method of *pratipakṣabhāvana*, as Patañjali describes, involves developing right type of thinking, which brings out the defects in the unwholesome thought. It follows the Buddhist idea of *pratipakṣabhāvanā* in spirit and content.

We have seen in the comments on the preceding aphorism, that Asaṅga talks about four types of veils of mind, namely distress (*paritamanā*), hindrances (*nīvaraṇa*), unwholesome thought (*vitarka*) and dogmatic adherence to the notion of *ātman* (*ātmasaṁpragraha*) and four means for removing or preventing them, namely *svabhāvaparijñāna*, *nidāna*, *ādīnavaprijñāna* and *pratipakṣabhāvanā*. After stating this, Asaṅga goes on to describe how each of these means are to be applied to these veils. Hence, while describing the application of *pratipakṣabhāvanā* to *vitarka* he says: "Antidote (*pratipakṣa*) in the case of the veils namely hindrances onward (that is, hindrances, unwholesome thought and dogmatic adherence to the notion of *ātman*) consists in cultivation of right mental attitude in contradistinction with wrong mental attitude".[87]

The other details of the "cultivation of the antidote" that Patañjali describes also reflect relevant Buddhist notions:

***kṛta-kārita-anumodita*:** In Buddhist *sūtras* we find the description of the three ways in which injury is made.[88]

***Lobha-krodha-moha-pūrvaka*:** In Buddhism the three passions, viz. desire, aversion and delusion (*lobha*, *dveṣa* and *moha*), are regarded as "unwholesome roots" (*akuśalamūla*). (AKB IV.68) Here Patañjali has only replaced "aversion" by "anger" ("*dveṣa*" by "*krodha*").

***Mṛdu-madhya-adhimātra*:** This three-fold classification is used in Buddhist Abhidharma in different contexts. Just as *indriya* (faculty) can be of three kinds, defilements also can be of three kinds. (AKB VI.33)

***Duḥkha-ajñāna-anantaphalāḥ*:** That *vitarkas* (in the sense of *kliṣṭavitarka* (defiled thoughts) cause immense suffering and *ajñāna* (ignorance, that is, wrong conception, misidentification) seems to be obvious in Buddhism.

अहिंसाप्रतिष्ठायां तत्सन्निधौ वैरत्यागः ।।35।।

Ahiṁsāpratiṣṭhāyāṃ tatsannidhau vairatyāgḥ.

When he is established in non-violence, there is abandonment of hostility in his presence.

Comments

Patañjali states in this aphorism that establishing oneself in non-injury leads the beings in the practitioner's vicinity to abandon enmity.

From this aphorism onwards up to aphorism II.45, Patañjali states one after the other the advantages of "establishing" (*pratiṣṭhā*) oneself in the five restraints and five observances. The source reference of the specific advantages Patañjali states are difficult to find. It is possible that Patañjali states them sometimes on the basis of his own experience, and sometimes on the basis of reasoning or imagination. The general idea, however, can be said to be rooted in the two theories of Buddhism: (1) the theory of supernormal powers which arise from perfection in meditation and (2) the Mahāyāna theory of ten planes (*bhūmi*) and special advantages of getting *established* in them.[89]

Though Patañjali's specific claim with reference to being established in non-injury seems to be exaggerated, the rational element in it is that when a person attains perfection in non-injury, his way of life influences other beings in the surrounding, so that they might develop an inclination to be non-injurious.

सत्यप्रतिष्ठायां क्रियाफलाश्रयत्वम् ।।36।।

Satyapratiṣṭhāyāṃ kriyāphalāśrayatvam.

When one becomes established in truthfulness, one's actions bear fruit.

Comments

Here Patañjali says that when a yogin gets established in truthfulness, his actions become fruitful/successful. This claim, too, like the claim in the preceding aphorism, appears to be an exaggeration. However, a rational justification of it is possible in the sense that a truth-speaking person is alert about understanding the facts as they are. Hence, his decisions are more realistic.[90]

अस्तेयप्रतिष्ठायां सर्वरत्नोपस्थानम् ।।37।।

Asteyapratiṣṭhāyāṁ sarvaratnopasthānam

When he is established in non-stealing, all the best things (literally: jewels) become available.

Comments

Here Patañjali is claiming that one who is established in non-stealing possesses all the jewels. This, like many other claims, is an exaggerated claim.[91]

ब्रह्मचर्यप्रतिष्ठायां वीर्यलाभः ।।38।।

Brahmacharyapratiṣṭhāyāṃ vīryalābhaḥ.

When he is established in celibacy, vigor is obtained.

Comment

Patañjali here is asserting what is commonly accepted as the advantage of celibacy, namely gaining and retaining virility.

अपरिग्रहस्थैर्ये जन्मकथंतासंबोधः ।।39।।

Aparigrahasthairye janmakathaṁtāsaṃbodhaḥ.

When he is established in non-possession there arises knowledge as to the "how" of one's (past, present and future) births.

Comment

Here Patañjali is linking non-possession with the knowledge of previous births. It is difficult to understand the link between the two. Maybe, the idea is that when one dissociates oneself from external belongings, one becomes clearly aware of one's internal features, including past and future experiences.

शौचात्स्वाङ्गजुगुप्सा परैरसंसर्गः ।।40।।

Śaucāt svāṅgajugupsā parair asaṁsargaḥ.

From cleanliness (arises) disgust for one's own body and disinclination towards contact with others.

Comments

As we have seen, Vyāsa distinguishes between external and internal cleanliness. While stating the advantages of cleanliness, Patañjali also seems to state the advantages of the two types of cleanliness differently. In this aphorism he is stating the advantages of bodily cleanliness, and in the next aphorism, those of mental cleanliness.

The idea of disgust towards one's own body (*svāṅgajugupsā*) is very much a part of the Buddhist approach to the human body. According to Buddhism such an attitude is developed through a special kind of meditation called contemplation on impurities (*aśubhabhāvanā*). Patañjali's theory of meditation does not make room for this kind of meditation. He regards this attitude as a result of perfection in cleanliness.

सत्त्वशुद्धिसौमनस्यैकाग्र्येन्द्रियजयात्मदर्शनयोग्यत्वानि च ।।41।।
Sattvaśuddhisaumanasyaikāgryendriyajayātmadarśanayogyatvāni ca.
(From cleanliness also result) purity of mind, joyfulness, one-pointedness, mastery over senses and fitness for vision of oneself.

Comments

Here Patañjali is talking about other advantages of "cleanliness", particularly those of mental cleanliness. Mental purity (*sattvaśuddhi*) is a general advantage, and many other advantages follow from that. The advantages, namely joyfulness (*saumanasya*), one-pointedness (*aikāgrya*) and mastery over senses (*indiyajaya*), are the notions common to Buddhism. Fitness to see *ātman* (*ātmadarśanayogyatva*) is unusual. Patañjali does not use such an expression elsewhere. The expression *ātmadarśana* can be interpreted in two ways: (1) "seeing oneself" (Here *ātman*, meaning "oneself", is taken as a reflexive pronoun and not as a noun.) and (2) "seeing the self" (Here *ātman* stands for the metaphysical self accepted by many Brahmanical systems.). The first interpretation is compatible with Buddhism, the second is not.[92]

सन्तोषादनुत्तमसुखलाभः ।।42।।
Santoṣād anuttamasukhalābhaḥ.
From contentment there is acquisition of extreme happiness.

Comments

Here Patañjali is talking about the advantage of (being established in) contentment. As we saw in the discussion of YS II.32, Vyāsa's definition of *santoṣa* (contentment) is close to the Buddhist concept of *santuṣṭi* (not desiring more than what one gets) at the background. Here Vyāsa identifies "perfection in contentment" with cessation of cravings (*tṛṣṇākṣaya*), and describes highest pleasure as the result of the latter. Vyāsa, in his commentary, quotes a verse, the source of which is likely to be in Buddhism.[93]

कायेन्द्रियसिद्धिरशुद्धिक्षयात्तपसः ।।43।।
Kāyendriyasiddhir aśuddhikṣayāt tapasaḥ.
From austerities is brought about the perfection of sense-organs and body by destroying (their) impurities.

Comments

Here Patañjali states the advantages of perfection in ascetic practice (*tapas*). They are attainment of supernormal bodily power and supernormal power of senses. The claims of such advantages are not typical of Buddhism. They can be found in Jaina and Vedic traditions of austerity as well.

स्वाध्यायाद् इष्टदेवतासम्प्रयोगः ।।44।।

Svādhyāyād iṣṭadevatāsamprayogaḥ

Due to self-study, one has direct association with one's desired deity.

Comments

Patañjali here is saying that due to study of scriptures, one attains the direct association with the desired deity. This idea is closely similar to the idea found in the Buddhist *sūtra*s where it is said that when *bodhisattva* studies Buddha's teachings from scriptures, he gets the revelation of thousands of the Buddhas.[94] Exaggerations apart, Patañjali seems to adopt this idea and generalize it by removing its sect-specific character.

This can also be taken to indicate the sectarian inclusiveness which is also found in some other aphorisms of Patañjali.[95]

समाधिसिद्धिरीश्वरप्रणिधानात् ।।45।।

Samādhisiddhir īśvarapraṇidhānāt.

From the resolve regarding the supreme being, results the expertise in meditative absorption.

Comments

As we have seen in the comments on the aphorisms concerning *īśvara* and *īśvarapraṇidhāna* (YS I.23–24), *īśvara* for Patañjali was the supreme being free from defilements, actions, fruition of actions and latent impressions, and *īśvarapraṇidhāna* can be understood as a resolve to become like *īśvara*. Here Patañjali is stating that achievement of (or expertise in) meditative absorption is the advantage of the resolve concerning *īśvara*.

स्थिरसुखमासनम् ।।46।।

Sthirasukham āsanam.

Posture means a steady and comfortable condition.

Comments

Patañjali is defining here the third limb of the eight-limbed Yoga called *āsana* ("posture").

We have to note here that there is a basic difference between *āsana* as understood in Pātañjala-yoga and as understood in Haṭhayoga. In Haṭhayoga we talk of different postures having different advantages for the functioning of different systems in our body. So *āsana*, as described in Haṭhayoga, is a technique for achieving and maintaining health and purification of channels in our body, the technique which is supposed to be indirectly helpful for purification of mind.

Patañjali defines *āsana* as "*sthirasukham āsanam*". Hence any "stable and comfortable" position of body can be *āsana*. Here the main purpose behind *āsana* is not health but being in a bodily and mental state which is convenient for meditation.

In early Buddhism there is no special focus on *āsanas*. Greater focus is on developing mindfulness in different natural positions. While sitting, standing, lying down or walking, one has to develop awareness and mindfulness. This is more important than choosing any particular posture. The images of the Buddha show that he, while meditating or preaching, generally used to sit in what is called *padmāsana*. But while teaching meditation, he is never found to be prescribing *padmāsana* or any other *āsana*.

Asaṅga, in *Śrāvakabhūmi*, gives some prescriptions regarding the sitting posture suitable for meditation. Asaṅga calls it *niṣaṇṇa*. He describes it as follows:

> One sits on a stage or a seat or grass-bed with crossed legs (*paryaṅka*) making the body straight and establishing mindfulness around the mouth (*pratimukhāṁ smṛtim upasthāpya*). This is called *niṣaṇṇa*.
>
> (*ŚB: 118)*

He then explains how one should remove hindrances in that position.

Patañjali's concept of *āsana* can be called a generalized and reformed version of this concept.

Surprisingly, Vyāsa interprets *sthirasukha* not as a definition of posture in general but as the name of one of the postures. He gives a list of thirteen postures out of which *sthirasukha* is one.[96] Vyāsa's interpretation of the aphorism is quite far-fetched and does not seem to be true to Patañjali's intentions.

प्रयत्नशैथिल्यानन्तसमापत्तिभ्याम् ।।47।।

Prayatnaśaithilyānantasamāpattibhyām.

(The posture is established/perfected) by relaxation of efforts and meditative attainment of the infinite.

Comments

In the preceding aphorism, Patañjali said that a posture is the one which is steady and comfortable. In this aphorism, he is stating how a steady and comfortable posture can be achieved. He states two means: (1) relaxing efforts or physical tension and (2) meditative attainment of an infinite object.

It is possible to interpret these two means as applicable to the two defining properties of *āsana* in their respective order. Hence the posture becomes steady (*sthira*) by avoiding exertion or artificial stretch of the body. And the posture becomes comfortable, because the mind focuses on an infinite (that is, a vast) object.

I didn't find the notion of relaxation of efforts (*prayatnaśaithilya*) in particular in the relevant Buddhist literature, though it is broadly consistent with the Buddhist approach opposed to self-mortification. Moreover, as we saw in the comments on the preceding aphorism, Asaṅga's concept of *niṣaṇṇa* does not contain the act of stretching of the body.

Similarly, Patañjali's notion of *samāpatti* seems to be inspired by the Buddhist theory of meditation (see the comments under YS I.41). The notion of *ananta-samāpatti* also can be traced to the latter. The objects of *samāpatti*, according to Vasubandhu, are eight, which include the spheres of infinite space (*ākāśānantyāyatana*) and infinite consciousness (*vijñānānantyāyatana*).[97] Possibly, Patañjali is suggesting that one can make the posture comfortable or peaceful if one practices meditative attainment on these infinities.

ततो द्वन्द्वानभिघातः ।।48।।

Tato dvandvānabhighātaḥ.

From that (i.e. perfection of the yogic posture) arises non-affection by opposite pairs.

Comments

Here Patañjali is stating the advantage of getting established in posture. The word *tataḥ* simply means "due to that", but from the Patañjali's use of the word *pratiṣṭhā* ("establishment") while stating the advantages of restraints (*yamas*) in aphorisms II.35–38, one can infer here, too, that Patañjali, by *tataḥ*, means "*āsanapratiṣṭhāyām*" (that is, when one becomes established in posture), which will also include mastering the technique of relaxation of efforts and meditation on a wide object. The advantage of such a practice, according to Patañjali, is that the opposite pairs, such as heat and cold, do not disturb the practitioner. There is no reference to this specific advantage in the Buddhist literature.

तस्मिन्सति श्वासप्रश्वासयोर्गतिविच्छेदः प्राणायामः ।।49।।

Tasmin sati śvāsapraśvāsayor gativicchedaḥ prāṇāyāmaḥ.

When the posture is (established), the distinct awareness (which one has) of the flows of inhalation and exhalation is called breath-regulation.

Comments

Just like *āsana*, *prāṇāyāma* is a point where Haṭhayoga and Pātañjala-yoga overlap. And just as Patañjali uses the term *āsana* in a sense different from that in which Haṭhayoga uses it, he probably uses the term *prāṇāyāma* also in a sense different from the Haṭhayoga sense of the term.

Patañjali defines *prāṇāyāma* as *gativiccheda* of inhalation and exhalation. Vyāsa interprets it as "absence of both",[98] which indicates destruction of the movement or stoppage of the movement of inhalation as well as exhalation, which is the gist of the Haṭhayogic concept of *prāṇāyāma*. But whether Patañjali meant this as the meaning of the term *prāṇāyāma* is really a question.

Prāṇāyāma in the Haṭhayogic sense is the method of purification of channels (*nāḍīs*) in the body, which ultimately leads to liberation through arousal of the dormant spiritual power (*kuṇḍalinī*). Patañjali's Yoga, on the other hand, centres on purification of mind. Patañjali could be interested in a breathing exercise to the extent to which it is related to purification of mind. Hence *prāṇāyāma* as a limb in Patañjali's Yoga could be interpreted differently.

In the Buddhist tradition we find reference to the two types of breath-exercises.

(1) Forcible stoppage of breathing: a kind of austere practice which the Gautama the *bodhisattva* tried in the initial stage of his pursuit of the right path to emancipation. It was called the practice in self-mortification (*duṣkaracaryā*). It is well-known that Gautama the *bodhisattva* performed different austere practices which included many Haṭhayogic breathing exercises.[99] However, *bodhisattva* Gautama stopped such exercises in self-mortification, realizing their futility, and discovered what he called the middle path (*madhyamamārga*). This middle path also had a scope for exercise with breathing. This brings us to the second type of breathing exercise.

(2) Development of the awareness of inbreathing and outbreathing, which is directly related to mental purification. It involves contemplation on breathing as it naturally takes place, but not an attempt to stop breathing.

It is possible that when Patañjali included *prāṇāyāma* as a part of his Yoga, he also meant something similar by this term. Hence, S. N. Tandon (1998: 90–91) interprets the term *gativiccheda* in Patañjali's definition of *prāṇāyāma* as "separation between movements". Tandon's claim can be supported by Patañjali's use of the word *vicchinna* in YS II.4. (The word *vicchinna* has the same root as *viccheda*: *vi+chid*.) The word *vicchinna* there means separated or manifested. It definitely does not mean "destroyed" or "stopped". It is relevant to note here that Vasubandhu in AKB uses the term *vicchinna* and *viccheda* in the senses of "separated" and "separation" (respectively) many times.[100] In the present context, *gativiccheda* can be interpreted as differentiation between or distinct awareness of the two motions (namely inhalation and exhalation).[101]

This brings Patañjali's *prāṇāyāma* closer to *ānāpānasmṛti* of Buddhism. This interpretation is also supported by the further description of *prāṇāyāma* that Patañjali gives in the next aphorism.

बाह्याभ्यन्तरस्तम्भवृत्तिर्देशकालसंख्याभिः परिदृष्टो दीर्घसूक्ष्मः ।।50।।
Bāhyābhyantarastambhavṛttir deśakālasaṁkhyābhiḥ paridṛṣṭo dīrghasūkṣmaḥ.

That (regulated breath), when it stays outside, inside or in the (middle) column, becomes long and subtle when observed fully in terms of its location, time and number.

Comments

Here Patañjali is explaining the breathing exercise further, which he calls *prāṇāyāma*, which, as we have seen, is close to "mindfulness of breathing" as discussed in Buddhism. Here Patañjali focuses on the observation of breathing in terms of its location, time and number (that is, frequency). Patañjali claims that through this type of exercise, the breath becomes long and subtle. This description of a breathing exercise has, as its background, the description of mindfulness of breathing (*ānāpānasmṛti*) as given by Asaṅga and Vasubandhu.

Vasubandhu (AKB VI.12d) talks about six aspects of *ānāpānasmṛti*, namely counting (*gaṇanā*), tracking (*anugama*), locating (*sthāpanā*), indicating (*upalakṣaṇā*), transforming (*vivarta*) and purifying (*pariśuddhi*). Apart from the aspects such as location and number, these aspects also include the ways of elevating the status of breath regulation to higher forms of meditation.

Asaṅga's description in *Śrāvakabhūmi* is quite elaborate. Asaṅga explains in detail how in *ānāpānasmṛti* the observation of breathing is to be conducted outside, inside and in the middle in terms of location, speed, counting and also as long and short, gross and subtle.

Outside, inside and in the middle

Asaṅga explains:

(1) *Āśvāsa* (inbreathing) is of two kinds—*āśvāsa* (inbreathing as such) and inbreathing in the interval (*antarāśvāsa*).[102] Inbreathing (as such) is the inward activity of air, which works up to the naval region. "Inbreathing in the interval" is the state of inbreathing when the inbreathing has stopped and outbreathing is yet to arise.[103]

(2) *Praśvāsa* (outbreathing) is of two kinds—*praśvāsa* (outbreathing as such) and outbreathing in the interval (*antarapraśvāsa*).[104] Outbreathing as such is the outward movement of air which begins from the naval region and goes out of mouth or nostrils.[105]

Hence, *āśvāsa* (inbreathing as such), which reaches the naval region, can be called *abhyantaravṛtti* (internally located) and *praśvāsa* (outbreathing as such), which reaches out of the mouth or nostrils, can be called *bāhyavṛtti* (externally located). And both *antarāśvāsa* (inbreathing in the interval) and *antarapraśvāsa* (outbreathing in the interval) can be called *stambhavṛtti* (located in the middle column).

Patañjali's expression *bāhyābhyantarastambhavṛtti*, in this way, seems to be an abridged version of Asaṅga's description.

Location, time and number

*Location (*deśa, bhūmi*)*

Asaṅga describes the locations of inbreathing and outbreathing in the following words:

> There are two types of locations of inbreathing and outbreathing. Which ones? The gross hole and the subtle hole. The gross hole ranges from naval region up to the openings of mouth and nose. And again, from the opening of mouth and nose up to the naval region. And subtle location consists of pores of the skin all over the body.[106]

*Time (*kāla*)*

Time factor refers to the speed of breathing: short and long. Asaṅga describes this aspect in the following words:

> When (the practitioner) observes breathing in (as such) or breathing out (as such), he learns "I am breathing in for long" or "I am breathing out for long". When he breathes in the interval or breathes out in the interval, he learns "I am breathing in for short in the interval" or "I am breathing out for short in the interval." That is because breathing in and out are long, and breathing in and out in the interval are short. He notes them as that; knows them as that.[107]

*Number (*saṁkhyā*)*

This factor refers to counting (*gaṇanā*). Asaṅga describes how counting is done in different ways in *ānāpāna* meditation. He refers to four types of counting:

(1) One to one counting (*ekaika-gaṇanā*): when he breathes in, he counts one, breathes out, he counts two and so on.
(2) Two to one counting (*dvayaika-gaṇanā*): when he completes one cycle of breathing in and out, he counts one; then having completed the cycle again he counts two and so on.
(3) Forward counting (*anuloma-gaṇanā*): he counts from one to ten and again starts from one.
(4) Backward counting (*pratiloma-gaṇanā*): starting from ten he counts backward up to one and again starts from ten.

(ŚB: 223–4)

Asaṅga adds that counting can be useful for concentration on breathing but it is prescribed to aspirants with feeble faculties (*mṛdvindriya*); aspirants with sharp faculties (*tīkṣṇendriya*) may be able to concentrate on breathing without counting. (ŚB: 225–6)

बाह्याभ्यन्तरविषयाक्षेपी चतुर्थः ।।51।।

Bāhyābhyantaraviṣayākṣepī caturthaḥ.

The fourth is that which sets aside external and the internal sphere.

Comments

Patañjali, in this aphorism, is introducing the fourth type of breath-regulation, which transcends inbreathing as well as outbreathing. Tandon (1998: 91) interprets the aphorism as:

> The fourth stage of *prāṇāyāma* casts aside the business of external and internal (breathing). In this stage the breath seems to come to a total stop—a phenomenon popularly known as *svataḥ kumbhaka*.

While explaining it further he says:

> When this process (i.e. the process of the breath being subtler and subtler) continues unabated and the respiration becomes subtle to the extreme, one may experience the fourth stage of *prāṇāyāma* where there is absolutely no movement of respiration either external or internal.
>
> *Tandon (1998: 92)*

As Tandon points out, the stoppage of breathing indicated by this fourth type of *prāṇāyāma* is a natural stoppage and not a forced one. Tandon's insight, I feel, can be appreciated.

However, Tandon (1998: 9, fn.1) identifies this state of stoppage of breath with the state called *saññāvedayitanirodha* ("cessation of perception and sensation"), in which the activities of speech, body and mind cease, which is the highest stage of concentration meditation. Such an identification between the highest type of breath-regulation and the highest stage of meditation seems to be far-fetched.

We find the description of such a natural stoppage of breathing in Asaṅga's formulation of the theory of *ānāpāna* (ŚB: 232). Asaṅga there explains the stage in which breathing naturally stops for some moments when the meditator experiences calmness (*praśrabdhi*). Asaṅga describes this state of cessation of both outbreathing and inbreathing in the following words:

> When outbreathing (as such) ceases, the outbreathing-in-the-interval also ceases, (but) inbreathing does not arise, then one observes a candid state (*sitām avasthām*) devoid of outbreathing and inbreathing. At that time, he breathes in bodily formations (*kāyasaṁskāra*) with calmness, learns that "I

am breathing in bodily formations with calmness." He breaths out bodily formations with calmness and learns like that.[108]

Here the idea of breathing in/out bodily formations with calmness probably refers to "bodily calmness" (*kāyikī praśrabdhi*) that one experiences in the first two stages of *dhyāna*.[109] It is interesting to note in this context that Vasubandhu refers to calmness (*praśrabdhi*) as a kind of pleasant air (*vāyu*) which arises due to a specific meditative concentration and spreads in the body.[110] So, the idea of the highest form of *prāṇāyāma* which results from the discussion is that in such a state, though breathing in and breathing out stops temporarily, one "breathes in and breathes out" the "air of calmness" which has spread all over the body.

ततः क्षीयते प्रकाशावरणम् ।।52।।

Tataḥ kṣīyate prakāśāvaraṇam.

By this the veil over illumination is thinned.

Comments

In this and the next aphorism Patañjali is stating the advantages of *prāṇāyāma*. One advantage according to him is that "the veil over illumination is thinned" (*prakāśāvaraṇa*). Vyāsa interprets the term *prakāśāvaraṇa* as "*vivekajñānāvaraṇīyaṁ karma*" (the karma which obscures discriminatory knowledge). This idea seems to be influenced by the Jaina concept of "*jñānāvaraṇīya-karma*" ("knowledge-covering karma").[111]

Patañjali's claim that the light-covering veil gets reduced through breath-regulation appears to be an exaggeration. Reduction of light-covering veil seems to be possible by elevating breath-regulation to the higher forms of meditation, as suggested by Vasubandhu.[112]

धारणासु च योग्यता मनसः ।।53।।

Dhāraṇāsu ca yogyatā manasaḥ.

And the mind acquires fitness for fixed attention.

Comments

Here Patañjali is stating the second advantage of *prāṇāyāma* practice. He is saying that by practicing breath-regulation, one becomes capable of fixing the mind on other objects. The significance of this can be understood on the background of the Buddhist theory of meditation. We have suggested that Patañjali's concept of *prāṇāyāma* can be better understood as keen observation of breathing as it naturally takes place rather than stopping or controlling breathing artificially. In this way, *prāṇāyāma* becomes an exercise in meditative concentration which is on par with the *ānāpāna* meditation of Buddhism.

In the Buddhist meditative practice, generally *ānāpāna* meditation is taught in the beginning so that the background is created for the meditative concentration on other objects.

This could be at the backdrop of the previously-mentioned claim of Patañjali.

स्वविषयासम्प्रयोगे चित्तस्य स्वरूपानुकार इवेन्द्रियाणां प्रत्याहारः ।।54।।

Svaviṣayāsamprayoge cittasya svarūpānukāra ivendriyāṇāṁ pratyāhāraḥ.

When the senses, separated from their corresponding objects, follow, as it were, the nature of mind, that is called withdrawal of senses (*pratyāhāra*).

Comments

Pratyāhāra literally means pulling back. The notion involves withdrawing senses from objects. This idea was implied in *Kaṭhopaniṣad* (3.4–6) where a person is compared with a charioteer who is supposed to control the horses symbolizing senses, which run after objects, with the help of the mind, which is like bridles.

Though in Buddhism we do not find the idea of *pratyāhāra* exactly in this form, the same function is performed by another term, *indriyasaṁvara* (protecting senses from objects).

The notion of *pratyāhāra* carries the imagery of pulling the senses back from the objects. The notion of *indriyasaṁvara* carries the image of closing the doors of senses for objects which try to enter through them.

Asaṅga in *Śrāvakabhūmi* explains *indriyasaṁvara* in detail. A person practicing *indriyasaṁvara* lives with "protected doors" with respect to sense organs. He protects the sense organ of mind (from defilements) and mind by sense organs and with the help of that he preserves his mindfulness.[113]

This practice is similar to *pratyāhāra* described by Patañjali.

ततः परमा वश्यतेन्द्रियाणाम् ।।55।।

Tataḥ paramā vaśyatendriyāṇām.

That brings supreme control over the sense organs.

Here Patañjali is stating the advantage of *pratyāhāra* (withdrawal of senses)—namely extreme control over senses.

As we have seen, *pratyāhāra* of Patañjali is comparable with *indriyasaṁvara* of Buddhism. The advantages of *indriyasaṁvara* stated by the Buddhists are similar to the advantage of *pratyāhāra* stated by Patañjali. The effects of *indriyasaṁvara* stated by Asaṅga are worth mentioning here:

> One lives with the doors protected by sense-organs, with protected mindfulness, wise/prudent mindfulness[114] etc. up to, he protects mind by mind.[115]

Notes

1 Patañjali uses the word in the sense of meditative absorption as he defines it in YS III.3.
2 "*rāgadveṣamohānāṁ* ca *tanutvāt*" (AKB 358, 370). Tandon (1998: 53) refers to a similar statement from *Dīghanikāya*: "*puna ca paraṁ mahāli bhikkhu tiṇṇaṁ saṁyojanānaṁ parikkhayā rāgadosamohānaṁ tanuttā sakadāgāmī hoti*".
3 The *Gītā* uses the term *kleśa* in this sense: "*kleśo'dhikataras teṣām avyaktāsaktacetasām/ avyaktā hi gatir duḥkhaṁ dehavadbhir avāpyate//* BG 12.5 [Those whose minds are attached to the non-manifest have greater pain (*kleśa*).]
4 Dhammajoti (2009: 324) quotes *Abhidharmāvatāra* in this context.
5 Some modern scholars translate the word as pain or affliction partly because they are not aware of the Buddhist background of the word *kleśa*.
6 "*samudayasatyaṁ katamat? kleśaḥ, kleśādhipateyaṁ ca karma*" (ADS: 43) [What is the truth called "arising (of suffering)"? Defilement and the action dominated by defilement.]
7 "*lakṣaṇaṁ katamat? Yo dharma utpadyamāno'praśāntalakṣaṇa utpadyamānena yena kāyacittap rabandhāpraśamapravṛttiḥ*" (ADS: 43).
8 "*ṣaṭ kleśāḥ katamat? rāgaḥ pratigho māno'avidyā vicikitsā dṛṣṭiś* ca" (ADS: 43).
9 "*ta eva dṛṣṭeḥ pañcākārabhedena daśa bahavanti*" (ADS: 43).
10 Asaṅga seems to distinguish between *kleśa* and *anuśaya* (or what he calls *kleśānuśaya* that is a dormant state of *kleśa*), as he describes how *kleśa* arises: "How does *kleśa* arise? When a dormant *kleśa* (*kleśānuśaya*) is not abandoned, some characteristic in the position of *kleśa* becomes apparent and there becomes present a wrong mental application. *Kleśa* arises in this way" ("*utthānaṁ katamat? . . . evaṁ kleśa utpadyate*" (ADS: 43). However, Vasubandhu does not seem to distinguish clearly between them. This is suggested by the statement about *anuśaya* in AK V.1a and the subsequent commentary (AKB V.1a) which is about *kleśa*, as if the two are one and the same. Dhammajoti (2009: 324) observes that the terms *anuśaya* and *kleśa* are the generic terms for defilement used in Abhidharma literature.
11 According to SK 47 there are five *viparyayas*. According to SK 48, the five *viparyayas* are *tamas* (of eight kinds), *moha* (of eight kinds), *mahāmoha* (of ten kinds), *tāmisra* (of eighteen kinds) and *andhatāmisra* (of eighteen kinds). All these varieties and sub-varieties are irrelevant to Patañjali's discussion of *kleśas* in YS. Also see comments on YS I.8.
12 The word *vicchinna* can be understood to mean "divided", "separated", "become manifest". It has the same root (*vi+chid*) as the word *viccheda*. See the comments on YS II.49.
13 Vasubandhu, in AK and AKB III.20–27, has discussed twelve-linked causal chain (*dvādaśanidāna*) in detail where he treats *avidyā, tṛṣṇā, upādāna* and *bhava* as *kleśas* and *avidyā* as the cause of other *kleśas*.
14 "*kleśāstribhirākārair veditavyāḥ. asti kleśo viparyāsamūlaḥ, asti viparyāsaḥ, asti viparyāsaniṣyandaḥ*" (YAB: 166).
15 Compare:

> **Vyāsa:** "*tatra kā* ***prasuptiḥ****? cetasi śaktimātrapratiṣṭhānām* ***bījabhāvo****pagamaḥ.* tasya ***prabodha*** *ālambane* ***sammukhībhāvaḥ***".
>
> *YB II.4*
>
> **Vasubandhu:** "*prasupto hi kleśo'nuśaya ucyate. prabuddhaḥ paryavasthānam. kā ca tasya* ***prasuptiḥ*** *asaṁmukhībhūtasya* ***bījabhāvā****nubandhaḥ. kaḥ* ***prabodhaḥ****?* ***saṁmukhiībhāvaḥ***".
>
> *AKB V.2a (p. 278)*

16 La Vallée-Poussin (1937) has acknowledged this. The four perversions are mentioned in AKB. "*catvāro viparyāsāḥ, anitye nityam iti, duḥkhe sukham iti, aśucau śucīti, anātmany ātmeti*" (AKB V.9).
17 "*viparyāsamūlam avidyā*" (YAB: 166).

18 Vasubandhu uses the term *asmitā* in the sense of *asmimāna*: "*nāsmitā dṛṣṭipuṣṭatvāt*" (AK V.11c).
19 "*asmimānaḥ katamaḥ? pañcasūpādānaskandheṣv ātmātmīyābhiniveśādyā cittasyonnatiḥ*" (ADS: 45).
20 "*buddhitaḥ paraṁ puruṣam ākāraśīlavidyādibhir vibhaktam apaśyan kuryāt tatrātmabuddhiṁ mohena*" (YB II.6).
21 For the meaning of the term *anuśaya*, see Dhammajoti (2009: 325–326).
22 The Buddhist Sanskrit term *anuśaya* seems to be based on the Pali word *anusaya*, which according to Pali English Dictionary means proclivity, bias, persistence of a latent or dormant tendency (always in bad sense). According to Edgerton's Buddhist Hybrid Sanskrit Dictionary, *anuśaya* means propensity (usually to evil), disposition to do something (usually evil).
23 "*sukhānuśayī tathā duḥkhānuśayīti anyeṣāṁ pāṭhaḥ. Sukhānuśayī rāgaḥ duḥkhānuśayī dveṣa iti te vyācakṣate*" (Vivaraṇa on YS(V) II.7).
24 "*sukhāyāṁ vedanāyām rāgo' nuśete, duḥkhāyāṁ pratighaḥ . . . ity uktaṁ sūtre*" (AKB II.3).
25 "*pratigho manyur jighāṁsā krodhaḥ sa dveṣaḥ*" (YB II.8).
26 "*ātmātmīyadhruvocchedanāstihīnāgradṛṣṭayaḥ/*
Ahetvamārge taddṛṣṭir etās tāḥ pañca dṛṣṭayaḥ//" (AK V.7).
27 "*śīlavratamātrakaṁ sāṅkhyayogādayaś ca na mārgo mokṣasya tāṁś ca mārgaṁ paśyati*" (AKB V.7).
28 "*samāsato hyete' nuśayāstraidhātukā darśana-prahātavyā bhāvanāprahātavyāś ca*" (In brief, these defilements belonging to all the three worlds are (two-fold): those to be abandoned by knowledge and those to be abandoned by meditation. (AKB V.3)).
29 For Vasubandhu's detailed account of the two types of *kleśas*, AKB V.4–12 (pp. 280–287) can be referred to.
30 "*nāvaśyavedanīyam aniyatam*" (AKB IV.50 (p. 229)).
31 "*tatra dṛṣṭadharmavedanīyam yatra janmani kṛtaṁ tatraiva vipacyate*" (AKB IV.50 (p. 230)).
32 "*upapadyavedanīyaṁ dvitīye janmani*" (AKB IV.50 (p. 230)).
33 "*aparaparyāyavedanīyaṁ tasmāt pareṇa*" (AKB IV.50 (p. 230)).
34 Vyāsa mentions *aniyatavipāka* in the commentary of the next aphorism; see YB II.12.
35 "*tatra nārakāṇām nāsti dṛṣṭajanmavedanīyaḥ karmāśayaḥ*" (YB II.12).
36 "*narakeṣu kuśalasya karmaṇas trividhasyākṣepo na dṛṣṭadharmavedaniyasya, tatra iṣṭavipākābhāvāt*" (AKB IV.51d).
37 "*kṣīṇakleṣānām api nasty adṛṣṭajanmavedanīyaḥ karmāśaya iti*" (YB II.12).
38 "*āryapudgalas tu yato vītarāgo naca pariāṇadharmā, sa tatropapadyavedanīyaṁ cāparaparyāyavedanīyaṁ ca karma na karoti*" (AKB IV.52c).
39 "*vipākahetur aśubhāḥ kuśalāś caiva sāsravāḥ*" (AK II.54).
40 As Asaṅga says, "With regard to the fruits such as family (that is, birth in a particular family), strength, appearance, life-span (*āyur*) and pleasant or painful experiences (*bhoga*), the major cause is good or bad action" ("*kulabalarūpāyurbhogādikasya tu phalasya prādhānyena śubhāśubhaṁ karma kāraṇam*") (YAB: 25).
41 "*ādyo jñānadarśanāvaraṇavedanīyamohanīyāyuṣkanāmagotrāntarāyāḥ*" (TAS 8.5) ["There are eight principal types of karmic bondage: knowledge-covering, intuition-covering, sensation, deluding, life-span, body, status and obstructive" (Tatia's translation)].
42 "*kṣemākṣemetarat karma kuśalākuśaletarat/puṇyāpuṇyam aniñjaṁ ca sukhavedyādi ca trayam//*" (AK IV.45).
43 "*punaḥ trīṇi. sukhavedanīyaṁ karma, duḥkhavedanīyam aduḥkhāsukhavedanīyaṁ ca*" (AKB IV.45cd).
44 La Vallée-Poussin (1937) has noticed this.
45 "*tisro hi duḥkhatā duḥkhaduḥkhatā, saṁskāraduḥkhatā vipariṇāmaduḥkhatā ca*" (AKB VI.3 (p. 329)).
46 "*sarve tu saṁskārāḥ saṁskāraduḥkhatayā duḥkhāḥ*" (AKB VI.3 (p. 329)).
47 "*kā punaḥ saṁskāraduḥkhatā? sukhānubhavāt sukhasaṁskārāśayo duḥkhānubhavād api duḥkhasaṁskārāśaya iti*" (YB II.15) [What is painfulness through impression? There is a

latent impression of pleasure due to pleasure-experience and there is latent impression of pain due to pain-experience.]

48 "*evaṁ etāni duḥkhāni akṣipātrakalpaṁ yoginam eva kliśnanti netaraṁ pratipattāram*" (YB II.15) [In this way these (three types of) sufferings inflict the yogin, who is like the eye-pot, and not another knower.]

49 "*yathā cikitsāśāstraṁ caturvyūhaṁ rogo rogahetur ārogyaṁ bhaṣajyam iti, evam idam api śāstraṁ caturvyūham eva, tad yathā saṁsāraḥ saṁsārahetur mokṣo mokṣopāya iti*" (YB II.15).

50 "*satyānyuktāni catvāri, duḥkhaṁ samudayas tathā/nirodho mārga ity eṣāṁ yathābhisamayaṁ kramaḥ//*" (AK VI.2).

51 "*satyānāṁ tu abhisamayānukūlā deśanā. kiṁ punaḥ kāraṇam evam eṣāṁ satyānām abhisamayaḥ? yatra hi sakto yena ca bādhyate yataś ca mokṣaṁ prārthayate tad evādau vyavacāraṇāvasthāyāṁ duḥkhasatyaṁ parīkṣyate. paścāt ko'sya hetur iti samudayasatyaṁ, ko'sya nirodha iti nirodhasatyaṁ, ko'sya mārga iti mārgasatyam.vyādhiṁ dṛṣṭvā tannidāna-kṣaya-bheṣajānveṣaṇavat. sūtre'py eṣa eva satyānāṁ dṛṣṭānto darśitaḥ. katamasmin sūtre? "caturbhir aṅgaiḥ samanvāgato bhiṣak talpasartety atra*" (AKB VI.2d). The last expression "*talpasartety atra*" in the passage is corrupt. I have accepted an alternative reading "*śalyāpahartety atra*" for translation.

52 "*duḥkhaṁ samudayo nirodho mārga iti. parijñeyaṁ praheyaṁ sākṣātkartavyaṁ bhāvayitavyam iti*" (AKB VI.54).

53 This comes under "*apratisaṅkhyānirodha*" as Vasubandhu defines it: "absolute obstacle to arising" ("*utpādātyantavighna*") (AK I.6cd).

54 "*tatra hātuḥ svarūpam upādeyaṁ vā heyaṁ vā na bhavitum arhati. Hāne tasyocchedavādaprasaṅgaḥ, upādāne ca hetuvādaḥ, ubhayapratyākhyāne śāśvatavāda ity etat samyagdarśanam*" (YB II.15).

55 "*yadi tarhi pudgalo nāsti, ka eṣa saṁsarati*?" (AKB IX: 471) ["If then there is no *pudgala*, who is this who transmigrates?"]

56 "*na cānyaḥ skandhebhyaḥ śakyate pratijñātuṁ śāśvataprasaṅgāt. Nāpy ananya uccheda-prasaṅgāt*" (AKB IX: 462) [Pudgala cannot be claimed to be different from the aggregates because this will lead to the problem of eternality. Nor can it be claimed to be identical with them because it will lead to the problem of annihilation.]

57 "*asty ātmetyānanda śāśvatāya paraiti, nasty ātmetyānandocchedāya paraiti*" (AKB IX: 470, 471).

58 "*smṛtiṁ tarhi kaḥ karoti? uktaḥ sa yas tāṁ karoti, smṛtihetuś cittaviśeṣaḥ*" (AKB IX: 472–73) [Who recollects then? We have stated who recollects; a special kind of mind, which is the cause (*hetu*) of recollection.] The whole first paragraph on page 473 explains how *citta* functions as cause.

59 "*smṛtijo hi chhandaḥ, chhandajo vitarko vitarkāt prayatnaḥ prayatnād vāyus tataḥ karmeti*" (AKB IX: 477).

60 "*saṁyogas tatkṛtas sargaḥ*" (SK 21).

61 "*ṣoḍaṣe ṣaṭ sarāgasya*" (AK VII.22a). Vasubandhu explains the meaning: at the time of the sixteenth moment of knowledge, the aspirant who still has attachment, has six types of knowledge.

62 "*vītarāgasya sapta tu*" (AK VII.22b).

63 "*sarāgabhāvanā marge tadūrdhvaṁ saptabhāvanā*" (AK VII.22cd).

64 "*kṣāntayo dṛṣṭir na jñānam*" (AKB VII.1a); "*anāsravā prajñā dṛṣṭiḥ jñānaṁ ca*" (AKB VII.1c); "*laukikī prajñā sarvaiva jñānam*" (AKB VII.1d).

65 "*sapta bhūmayaḥ catvārī dhyānāni trayaścārūpyāḥ, tāsāṁ jayo vairāgyaṁ tasmin saptabhūmike vairāgye . . . sapta jñānāni bhāvyante*" (AKB VII.23).

66 There is a little difference between *kṣayajñāna* (knowledge of destruction) and *anutpādajñāna* (knowledge of non-origination). *Kṣayajñāna* is of the form "What is to be done has been done" and *anutpādajñāna* is of the form "Nothing remains to be done" (AKB VII.7). Vyāsa uses both the forms in his explanation of first four types of *prajñā*.

67 "*dve vimuktī sūtra ukte—cetovimuktiḥ prajñāvimuktiś ca*" (AKB VI.76c).

68 La Vallée-Poussin (1937) remarks that the last three forms of *prajñā* of Vyāsa's classification match with *anutpādajñāna* or *kṣayajñāna* held by the *arhat*. This is not convincing. Their affinity to Sāṅkhya is more vivid.

69 ŚB: 261. Asaṅga discusses the three-fold training of a yogin in *Śrāvakabhūmi* (ŚB: 261–65). He concludes with the statement, "*tatra yoginā yogaprayuktena śikṣitavyam*" (ŚB: 265) [A yogin, who is engaged in Yoga should be trained in it.]

70 "*kā punar āsāṁ śikṣāṇām ānupūrvī? [su]viśuddhaśīlasya [a]vipratisāraḥ. avipratisāriṇaḥ prāmodyaṁ, prītiḥ, praśrabdhiḥ sukhaṁ, sukhitasya cittasamādhiḥ. Samāhitacitto yathābhūtaṁ prajānāti . . . evam imāni śīlāni bhāvitāni agratāyām upanayanti yadutānupādāya parinirvāṇam iti iyam āsāṁ śikṣāṇām ānupūrvī*" (ŚB: 263–4). (Emendations are suggested in brackets according to the consistency of meaning.)

71 For a detailed account and discussion, see my paper "Re-understanding Indian Moral Thought" (Gokhale, 2002).

72 Vyāsa says, "The other restraints succeeding *ahiṁsā* are rooted in *ahiṁsā* and are stated for the establishment of *ahiṁsā*" (YB II.30) (translation mine). For my discussion of how Jaina thinkers regard non-violence as the highest value, see Gokhale (2009).

73 "*deśasarvato'ṇumahatī*" (TAS 7.2) ["Partial abstinence is a small vow and complete abstinence is a great vow" (Tatia's translation)].

74 "*aṇuvrato'gārī*" (TAS 7.15).

75 "*sannihitasādhanād adhikasyānupāditsā*" (YB II.32).

76 "*labdheṣu kila praṇīteṣu cīvarādiṣu bhūyaskāmatā asantuṣṭiḥ*" (AK VI.6) (when one gets excellent robes etc.; desiring them more is non-contentedness).

77 Dhammapada praises *santuṭṭhi* (=*santoṣa*) as the highest wealth ("*paramaṁ dhanaṁ*"), *Dhammapada*, 15.8.

78 "*Buddhānāṁ tapaḥ sukham*" (AKB I.10d).

79 "*na cāryāḥ santaḥ kaṣṭāni tapāṁsi tapyeran*" (AKB III.94c).

80 "*rātryāḥ paścime yāme laghu laghv eva prativibudhya bhāṣye vā svādhyāyakriyāyāṁ vā yogaḥ karaṇīyaḥ*" (ŚB: 126).

81 Thanks to Professor Mathrimurthi for suggesting this translation of *pratipakṣa*.

82 The expression *vitarkabādhane* can be interpreted in two ways: (1) as temporal locative (*sati-saptamī*) which means "when unwholesome thoughts trouble (the Yogin)" or (2) as meaning "with respect to obstruction/removal of unwholesome thoughts".

83 See the comments on YS I.17.

84 "*tatra vitarkaḥ kāmavitarkādayaḥ kliṣṭā vitarkāḥ*" (ŚB: 399).

85 Only with the difference that Patañjali has used the word *pratipakṣabhāvana* (neuter gender) instead of *pratipakṣabhāvanā* (feminine gender) used by the Buddhist thinkers.

86 In Pali texts they are found mentioned as *kāma*, *vihiṁsā* and *vyāpāda*. See "*vitakka*" in Davids and Stede (1997): Pali English Dictionary.

87 "*nivaraṇādīnāṁ ayoniśomanasikāraviparyayeṇa yoniśomanasikārabhāvanā pratipakṣo veditavyaḥ*" (ŚB: 401).

88 As Tandon (1998: 13) translates a quote from Anguttara 3.7.1, "Monks, one possessed of three qualities is put into Hell according to his deserts. What three? One is himself taker of life, encourages another to do the same and approves thereof".

89 It is interesting to note the similarity with regard to the use of the notion of "establishment/established" (*pratiṣṭhā*, *pratiṣṭhita*) in *Yogasūtra* on the one hand and the Buddhist texts on ten *bhūmi*s on the other. Patañjali uses the word *pratiṣṭhā* in four aphorisms (YS II.35–38) and *sthairya* (stability) in one aphorism (II.39). And though in another five aphorisms he does not use any such word, he does so probably only for avoiding repetition. The language of *pratiṣṭhā* indicates becoming established in the given moral-spiritual virtue. The *Daśabhūmīśvara-sūtra* of Mahāyāna Buddhism in its theory of ten *bhūmi*s also uses the language of being established in a certain plane (*bhūmi*). While describing every plane, the *sūtra* describes the advantages of being established (*pratiṣṭhita*) in the respective plane. (See, for instance, DBS: 37, 45, 61, 74, 87, 105, 125, 160, 193.) Just as the Mahāyāna author states the advantages of being established in a plane in extreme or exaggerated language, similarly Patañjali states advantages of being established in restraints and observances, sometimes in extreme or exaggerated language.

90 In this context, Tandon (1998: 20) refers to what is called *saccakiriyā* in the Pali Buddhism (Skt: *satyakriyā*). It is suggested from the examples he has quoted that one who establishes oneself in truthfulness can use his mastery as a means for achieving a required result. It is possible that Patañjali was inspired by such beliefs when he made these claims.

91 Professor Mathrimurthi brought to my notice in this regard that the Buddhist texts have frequently emphasized that when the yogin is established in non-stealing he never loses anything and is wealthy as the karmic result of his meritorious act.

92 One could suspect that Patañjali had an ambivalent approach to *ātman*. In this context, see comments on YS IV.25.

93 "*yac ca kāmasukhaṁ loke yac ca divyaṁ mahat sukham /tṛṣṇākṣayasukhasyaite nārhataḥ ṣoḍaśīaṁ kalām //*" As pointed out by La Vallée-Poussin (1937), the original Pali version is found in *Nettippakaraṇa* (165). Its Sanskrit projection can be seen in UV XXX.31: "*yac ca kāmasukhaṁ loke yac cāpi divijaṁ sukhaṁ /tṛṣṇākṣayasukhasyaitat kalāṁ nārghati ṣoḍaśīm//*". The Sanskrit verse with some variation is found in several places in *Mahābhārata* also (for example, *Mokṣadharma*, 168.37, 171.51, 268.06); however, its source is likely to be Buddhist.

94 According to *Daśabhūmīśvara-sūtra*, bodhisattva on the very first plane, namely *pramuditā-bhūmi*, studies *śāstras* and the world, and develops various spiritual properties due to which many Buddhas directly appear to him in their gross form as well as through his resolve (DBI: 25–6).

95 See the comments on YS I.39 and II.32.

96 "*tad yathā* (1) *padmāsanaṁ* (2) *vīrāsanaṁ* (3) *bhadrāsanaṁ* (4) *svastikaṁ* (5) *daṇḍāsanam* (6) *sopāśrayaṁ* (7) *paryaṅkaṁ* (8) *krauñcaniṣadanaṁ*, (9) *hastiniṣadanaṁ* (10) *uṣṭraniṣadanaṁ* (11) *samasaṁsthānaṁ* (12) *sthirasukhaṁ* (13) *yathāsukhaṁ cety evamādīni*" (YB II.46).

97 "*ity etāny aṣṭau maulāni samāpattidravyāṇi yaduta catvāri dhyānāni catvāra ārūpyā iti*" (AKB VIII.5ab). ["These are the basic eight substances of attainment: The four meditations and the four formless spheres".] The four formless spheres are infinite space, infinite consciousness, nothingness and neither perception nor non-perception (AKB VIII.4).

98 "*ubhayābhāvaḥ*" (YB II.49).

99 *Mahāsaccakasutta* records Buddha's narration of his own practices: "I checked inhalation and exhalation from mouth, nostrils and ears. As I suppressed my breathing thus, a tremendous burning pervaded my body. Just as if two strong men were each to seize a weaker man by his arms and scorch and thoroughly burn him in a pit of glowing charcoal, even so did a severe burning pervade my body" Thomas (2005: 65).

100 For example, "*dūrāntaravicchinnaṁ vyutthānacittam*" ["The outward-directed mind which is remotely separated (from the meditating mind)"] (AKB II.62ab, p. 100). Also see the uses of the words *viccheda* and *vicchinna* in AKB III.10–11 where Vasubandhu discusses whether birth-mind (*upapattibhavacitta*) follows the death-mind (*maraṇabhavacitta*) without separation (*aviccheda*).

101 Asaṅga makes the distinction between the two motions as follows, "*dve āśvāsapraśvāsayor gatī. Katame dve? Āśvāsayor adhogatiḥ, praśvāsayor ūrdhvagatiḥ*" (ŚB: 221). ["Inbreathing and outbreathing have two (different) motions. Which two? Inbreathing has downward motion; outbreathing has upward motion".]

102 "*tatra dvāv āśvāsau, katamau dvau? āśvāso' ntarāśvāsaś ca*" (ŚB: 220). Here the word *antarāśvāsa* is to be derived as "*antare āśvāsaḥ*", which means inbreathing in the gap or interval (*antara*).

103 "*tatra āśvāsaḥ yaḥ praśvāsasamanantaram antarmukho vāyuḥ pravartate. yāvan nābhīpradeśāt*" ibid.

104 "*dvau praśvāsau. katamau dvau? praśvaso' ntarapraśvāsaś ca*" (ŚB: 220).

105 "*bahirmukho vāyuḥ pravartate. bahir nābhīpradeśam upādāya yāvan mukhān nāsikāgrāt tato vā bahiḥ*" (ŚB: 220).

106 "*dve āśvāsapraśvāsayor bhūmī, katame dve? audārikaṁ ca sauṣiryaṁ sūkṣmaṁ ca sauṣiryam. tatraudārikaṁ sauṣiryaṁ nābhīpradeśam upādāya yāvan mukhanāsikādvāram mukhanāsikādvāram upādāya yāvannābhīpradeśasauṣiryam. sūkṣma-sauṣiryaṁ katamat? sarvakāyagatāni romakūpāni*" (ŚB: 221–222).

107 "*yadā āśvāsaṁ vā praśvāsaṁ vā ālambate tadā dīrgham āśvasimi praśvsimīti śikṣate. yadā antarāśvāsam antarā praśvāsaṁ vā ālambanīkaroti tadā hrasvam āśvasimi praśvasimi iti śikṣate. tathā hi āśvāsapraśvāsā dīrghāḥ pravartante, antarāśvāsā antarapraśvāsāś ca hrasvās te tathaiva pravartante. tathaiva upalakṣayati, jānāti*" (ŚB: 232).

108 "*niruddhe ca praśvāse antarapraśvāse ca anutpanne āśvāse'ntarāśvāse ca praśvāsāśvāsaśūnyāṁ tadvyupetāṁ tadvyavahitāṁ sitām avasthām ālambanīkaroti. tasmin samaye praśrabhya kāyasaṁskārān āśvasan praśrabhya kāyasaṁskārān āśvasimīti śikṣate . . . praśvasimīti śikṣate*" (ŚB: 232).

109 "*|praśrabdhisukham ādyayoḥ*" (AK VIII.9b).

110 "*samāpannasya kathaṁ kāyavijñānam iti cet. samādhiviśeṣajena prasrabdhisaṁjñakena sukhavedanīyena vāyunā kāyaspharaṇāt*" (AKB VIII.9b) [If you ask, "How can there be bodily consciousness, when one attains a meditative state?", the answer is, "(This happens) because a pleasant air called 'calmness', which arises from a special type of meditative absorption spreads all over the body"].

111 As Tatia (2007) translates TAS 8.4, "There are eight principle types of karmic bondage: knowledge-covering, intuition-covering, sensation, deluding, lifespan, body, status, and obstructive".

112 AKB VI.12d. See the comments on YS II.50.

113 "*yathāpīhaikatya indriyair guptadvāro viharati ārakṣitasmṛtir nipakasmṛtir iti . . . yāvad rakṣati mana indriyaṁ mana indriyeṇa*" (ŚB: 63–64).

114 "*nipakasmṛti*", for *nipaka*, see Edgerton's Buddhist Hybrid Sanskrit Dictionary.

115 "*indriyair guptadvāro viharati, ārakṣitasmṛtir nipakasmṛtir iti vistaraḥ . . . yāvad rakṣati mana indriyaṁ mana indriyeṇa*" (ŚB: 64).

3

ON SUPERNORMAL POWERS (*VIBHŪTIPĀDAḤ*)

In this third chapter, which consists of fifty-five aphorisms, called *Vibhūtipādaḥ*, Patañjali elaborates on the different supernormal powers that a yogin, according to him, can attain by using the technique of integrated concentration (*saṁyama*). Here is an aphorism-wise summary of the third chapter of the *Yogasūtra*.

(1–8:) Patañjali begins the chapter by resuming the discussion of eight-limbed yoga, which he left incomplete at the fifth limb in the second chapter. The last three limbs he defines here are fixed attention (*dhāraṇā*), meditation (*dhyāna*) and meditative absorption (*samādhi*). He refers to these three limbs as the internal aspect of yoga as compared to the earlier five limbs, which he calls the external aspect. He asserts that these three are still external as compared to the seedless meditative absorption. He defines integrated concentration (*saṁyama*) as the last three limbs directed to a common object. **(9–16:)** Patañjali then talks about the three-fold transformation that takes place in a thing when the thing changes: change of the property, temporal characteristic and state of existence. He explains how such changes take place in the mind when one enters the state of meditative absorption. He claims that through the integrated concentration on three-fold transformation, the yogin can know the past and future of a thing. **(17–24:)** Patañjali then introduces the supernormal powers, such as the knowledge of the utterances of all beings, the knowledge of previous births, that of other minds, the power of disappearance, the pre-knowledge of death and the powers of different kinds including elephant power. **(25–37:)** Patañjali continues by introducing supernormal knowledges, such as the knowledge of subtle, hidden and remote objects; of the cosmos; of the stars and their movements; and the structure of the body. He states that control over hunger and thirst and steadiness of body are possible through integrated concentration. He claims that extrasensory perception and even omniscience is possible by this method. He warns, however, that though achievements in the extravert state of mind, these are all obstacles in meditative trance. **(38–46:)** Patañjali continues the story of supernormal powers. He claims that a yogin's mind, through integrated concentration, can enter in someone else's body; he can travel through water, mud or thorns without these things sticking to him; he can radiate, attain divine ear or move through space; can become extremely small, large, light or heavy; can achieve a beautiful and strong body. **(47–55:)** In the last few aphorisms of the chapter, Patañjali talks about supernormal powers conducive to liberation, such as conquest of sense organs, that of primordial nature, discriminating knowledge and isolation.

In each case, Patañjali mentions a suitable object of concentration, by integrated concentration on which the yogin achieves the respective supernormal knowledge or power.

The comments on these aphorisms will reveal how Patañjali's claims about supernormal knowledges and powers largely reflect the claims about *abhijñā*s and *ṛddhi*s made by the Buddhist tradition as represented by Asaṅga and Vasubandhu.

देशबन्धश्चित्तस्य धारणा ||1||

Deśabandhaś cittasya dhāraṇā.

Fixed attention (*dhāraṇā*) is fixing the mind on a (particular) location.

Comments

After *Pratyāhāra* Patañjali introduces three limbs, all related to meditation, which he calls *antaraṅgayoga*. They are *dhāraṇā*, *dhyāna* and *samādhi*.

The first one is *dhāraṇā*. Patañjali defines *dhāraṇā* as "Fixing the mind on a particular location". According to Vyāsa this location is generally somewhere in the body, though he also includes external locations.[1]

As I have noted in Appendix I, *dhāraṇā* of Patañjali can be approached in the Buddhist context in terms of *savitarka-dhyāna*. In early Buddhism, form-meditation (*rūpadhyāna*1) is classified into four kinds (I use Sanskrit terminology for the sake of convenience):

(1) *Dhyāna* with *vitarka-vicāra*, *prīti*, *sukha* and *ekāgratā*
(2) *Dhyāna* with *prīti*, *sukha* and *ekāgratā*
(3) *Dhyāna* with *sukha* and *ekāgratā*
(4) *Dhyāna* with *ekāgratā*

Out of them the first *dhyāna* is divided into two stages—(1a) *dhyāna* with *vitarka* and *vicāra* and (1b) *dhyāna* with *vicāra* but without *vitarka*. Here I want to suggest that the understanding of *vitarka* in Pali Buddhism brings it close to the concept of *dhāraṇā*. Rhys Davids and Stede in their Pali English Dictionary (p. 688) understand *vitakka* (Sanskrit: *vitarka*) in the context of first meditative trance as "initial application", or "the directing of concomitant properties towards the object" and *vicāra* as "the continued exercise of the mind on the object". By referring to simile used in *Dīghanikāya* II.277, they also say that the description there gives *vitakka* the characteristic of fixity and steadiness; *vicāra* that of movement and display. Here the very first *dhyāna* (that is, "1a") is close to what Patañjali calls *dhāraṇā*. The next stage in which there is no *vitarka* but only *vicāra* (that is, "1b") is close to *dhyāna*, and later stages in which *ekāgratā* (*samādhi*) marked by absorption of mind in the object is important, are close to *samādhi*.

Another indicator of the notion of *dhāraṇā* can be found in Vasubandhu's notion of *cittadhāraṇa*. According to Vasubandhu's explanation, the initial step in the Buddhist meditative practice is either *ānāpāna* (awareness of inbreathing and outbreathing) or *aśubhabhāvanā* (concentration on impurities). We have already talked about *ānāpāna*. *Aśubhabhāvanā* is practiced by fixing the mind on different parts of the body, such as blood, flesh and skeleton with the realization of their impurity. Having fixed the mind part by part, finally one fixes the mind on the middle of the eyebrows, "*bhrūmadhye cittadhāraṇāt*" (AK V.11). The use of the word *dhāraṇā* here is significant. Though in Patañjali's theory of meditation, the concept of *dhāraṇā* does not have the context of "meditation on impurities", fixing the mind on a location in the body is common between the two conceptions of *dhāraṇā*.

तत्र प्रत्ययैकतानता ध्यानम् ।।2।।

Tatra pratyayaikatānatā dhyānam.

The continuous flow of the same cognition therein is meditation (*dhyāna*).

Comments

As suggested in the comments on the previous aphorism, *dhāraṇā* can be understood as the meditative state with initial application of mind to the object, *dhyāna* can be understood as the meditative state with sustained application. This sustained and uninterrupted flow of mind can also be accompanied by *prīti* and *sukha* (joy and happiness). What is most important in *samādhi*, the next stage, is the concentration (one-pointedness) of mind on the object (*ekāgratā*).

तदेवार्थमात्रनिर्भासं स्वरूपशून्यमिव समाधिः ।।3।।

Tad evārthamātranirbhāsaṁ svarūpaśūnyam iva samādhiḥ.

The same, when it manifests the object alone and is devoid, as it were, of itself (i.e. its own nature) is the meditative absorption (*samādhi*).

Comments

According to Buddhist theory of meditation, *samādhi* means one-pointedness of mind, or concentration of mind. Accordingly, *samādhi* is not prominent in the first *dhyāna*. It becomes more and more prominent in the second *dhyāna* onwards. That is why they say that in the first *dhyāna* there are joy and happiness (*prītisukham*) caused by seclusion (*vivekajaṁ prītisukham*); in the second *dhyāna*, however, in which there is no *vitarka* or *vicāra*, there are joy and happiness caused by concentration (*samādhijaṁ prītisukham*) with the development of *samādhi*, one becomes free from joy, happiness, and develops a sort of indifference/detachment (*upekṣā*). Patañjali's definition of *samādhi* reflects the concentration aspect as well as detachment aspect of *samādhi*:

(1) ***Arthamātranirbhāsaṃ***—where only the object appears. This indicates the concentration aspect of *samādhi*, according to which the mind gets absorbed into the object.
(2) ***Svarūpaśūnyam iva***—There is no "I-notion" present. This indicates the *upekṣā* (disinterestedness, detachment) aspect of *samādhi*.[2]

त्रयमेकत्र संयमः ||4||

Trayam ekatra saṁyamaḥ.

The three directed to the same object is integrated concentration (*saṁyama*).

Comments

We have seen that *dhāraṇā*, *dhyāna* and *samādhi* can be correlated with different stages in the Buddhist scheme of four *dhyānas*. *Dhāraṇā* can be correlated with the first *dhyāna* with domination of *vitarka* (initial application); *dhyāna* can be correlated with the intermediate *dhyāna* ("*dhyānāntara*") accompanied by *vicāra*; and the other *dhyānas* (second, third and fourth) can be considered as the higher and higher forms of *samādhi*. Patañjali is here introducing the concept of *saṁyama*, which he defines as all three (*dhāraṇā*, *dhyāna* and *samādhi*) in togetherness. In Appendix II, I have shown how, according to Asaṅga, with the pursuit of pure *dhyāna* with its different stages along with the other repeated practices (*bahulīkāra*) such as mental application (*manasikāra*), accurate knowledge (*pratisaṁvit*) and resolve (*adhimokṣa*), one attains various supernormal cognitions and powers (*abhijñā* and *ṛddhi*). Such an idea of an enriched and deepened form of concentration seems to be at the background of Patañjali's concept of *saṃyama*. Patañjali does not explain such a complex method of concentration. He presents a simplified version of the method, namely "triple concentration".

तज्जयात्प्रज्ञालोकः ||5||

tajjayāt prajñālokḥ.

With its mastery, comes the light of wisdom.

Comments

In the Buddhist theory of meditation, after the second *dhyāna*, there arise *samprajñāna*, *smṛti* and *upekṣā*. Vasubandhu terms *samprajñāna* as *prajñā*.[3] In the first chapter also Patañjali referred to *ṛtambharā prajñā* arising after the "*adhyātma-prasāda*" (internal serenity). That is exactly after second *dhyāna*, according to the Buddhist classification. Hence, the passage from *samādhi* to *prajñā* in Chapter 1 and Chapter 3 go parallel to each other. Of course, there is one difference. In the first chapter, Patañjali introduced the highest form of *prañā*, namely *ṛtambharā prajñā*, because the main theme was final liberating trance, namely seedless *samādhi*. In this chapter, the main theme is supernormal cognitions and powers. So here the "light of wisdom" (*prajñāloka*), which he is introducing, is more general, which can be applied to different planes.

तस्य भूमिषु विनियोगः ।।6।।

Tasya bhūmiṣu viniyogaḥ.

It (=the light of wisdom) is applied to (different) planes.

Comments

That wisdom (*prajñā*) is to be applied to many planes (*bhūmis*) is commonplace in the Buddhist theory of meditation. For example, Vasubandhu says: "The mindfulness of breathing is a kind of wisdom, which can belong to five planes".[4] This also suggests that a certain type of insight (wisdom, knowledge) will be available in certain types of planes and not necessarily in all planes.

Buddhists accept a hierarchy of worlds, namely *kāmadhātu*, *rūpadhātu* and *arūpadhātu* (the worlds of sensuousness, form and formlessness, respectively). Through practice of meditation in *kāmadhātu*, a certain insight is available which will stimulate the practitioner to proceed to *rūpadhātu*, where certain aspects of *kāmadhātu* are not available. Again, through practice of meditation in *rūpadhātu*, a certain insight is available which will stimulate the practitioner to proceed to *arūpadhātu*, where certain aspects of *rūpadhātu* are not available.

Similarly, Vasubandhu tells a general principle about the higher forms of knowledge (and power), that is that they can take objects belonging to their own planes or the lower planes but not higher planes.[5]

त्रयमन्तरङ्गं पूर्वेभ्यः ।।7।।

Trayam antaraṅgaṃ pūrvebhyaḥ.

These three (i.e. fixed attention, meditation and meditative absorption) are internal (aspects of yoga) in comparison to the preceding limbs.

Comments

Yama and *niyama* are related mainly to regulation of conduct. *Āsana* is regulation of bodily position. *Prāṇāyāma* is concerned with regulation of breathing. *Pratyāhāra* is concerned with regulation of sense-organs. Now, all these aspects, namely bodily behavior, bodily position, breathing and sense-organs, which are regulated by the first five limbs of Yoga, are external to mind. *Dhāraṇā*, *dhyāna* and *samādhi*, on the other hand, are concerned with regulation of the mind itself.

This is the reason why Patañjali regards the earlier five limbs of Yoga as external limbs and the last three as the internal limbs of *yoga*.

तदपि बहिरङ्गं निर्बीजस्य ।।8।।

Tad api bahiraṅgaṃ nirbījasya.

That also is the external aspect with respect to seedless (meditative absorption).

Comments

The meditative states, named *dhāraṇā*, *dhyāna* and *samādhi*, are basically *samprajñāta* types of meditations; they are also *sabīja* (containing seeds of inflows). According to Buddhism objects and inflows (impurities, *mala*) are external to mind. Mind by its essence is shining (*prakṛti-prabhāsvara*) and pure. In *nirbīja-samādhi*, the mind shines in its original nature. For this reason, it seems, Patañjali calls the three stages of meditation as external (*bahiraṅga*) to seedless meditative absorption.

व्युत्थाननिरोधसंस्कारयोरभिभवप्रादुर्भावौ निरोधक्षणचित्तान्वयो निरोधपरिणामः ।।9।।

vyutthānanirodhasaṅskārayor abhibhavaprādurbhāvau nirodhakṣaṇacittānvayo nirodhapariṇāmaḥ.

The impressions of outward directedness are overpowered and those of cessation emerge; there is association of mind with the moment of cessation. This is cessation-type transformation (of mind).

Comments

In this and the next three aphorisms, Patañjali is giving three different descriptions of the change that takes place in the mind (*citta*) when it leaves its ordinary state and enters into meditative trance. They are in terms of transformation of a property (*dharmapariṇāma*), transformation of temporal dimension (*lakṣaṇapariṇāma*) and transformation of the state of thing (*avasthāpariṇāma*) respectively, as Patañjali himself will suggest in YS III.13. Hence, what he calls cessation-type transformation (*nirodha-pariṇāma*) here is the same as what he calls transformation of property (*dharmapariṇāma*) in YS III.13. Patañjali wants to say here that the change in mind which takes place at the time of entrance into meditative trance can be described as a change of properties. Outward-directedness is the property which goes, and the new property, namely "cessation (of mental states)", arises. The property-bearer, namely mind, remains intact in this process of transformation.

Patañjali's above explanation can be understood as response to the Buddhist theory of meditation according to which the mind along with the mental state (*citta* along with *caitta*) changes every moment such that the earlier mind is supposed to be the "immediate cause" (*samanantarapratyaya*) of the next mind.[6] According to this theory, the extravert mind (*vyutthāna-citta* or *vyutthita-citta*) is a different mind, and the cessation-type mind or cessation-attaining mind (*nirodha-samāpatti-citta*) is a different mind. When one enters meditative trance, the extravert mind becomes the immediate cause of the cessation-attaining mind and when one comes out of trance, the cessation-attaining mind becomes the immediate cause of the extravert mind.[7]

Patañjali, in response to this Buddhist theory is saying that *vyutthāna* and *nirodha* are not two different minds but two properties which the same mind can have.

तस्य प्रशान्तवाहिता संस्कारात् ।।10।।

Tasya praśāntavāhitā saṁskārāt.

That (=the mind in the moment of cessation) has a peaceful flow due to the latent impressions (of cessation).

Comments

This is very much an echo of what Vasubandhu says about the person who comes out of the state of *nirodha-samāpatti*.

> The person who comes out of the attainment (of cessation) experiences extreme peace of mind. That is because this attainment is similar to *nirvāṇa*.[8]

सर्वार्थतैकाग्रतयोः क्षयोदयौ चित्तस्य समाधिपरिणामः ।।11।।

Sarvārdhataikāgratayoḥ kṣayodayau cittasya samādhipariṇāmaḥ.

The decay of dispersiveness (literally: attention to everything) and rise of one-pointedness of mind are the meditative absorption-type of transformation of mind.

Comments

The same transformation of mind which was described in the earlier aphorism, in terms of cessation, can be described in terms of meditative absorption. Now, what happens when meditative absorption takes place? The mind was initially about everything (*sarvārthatā*) (It could have anything as its object) but this about-everything-ness gets destroyed and single-pointedness arises. This is the impact of meditative absorption.

If there is one to one correlation between the three types of transformations described in the three aphorisms (YS III.9, 11 and 12) and the three kinds of transformations mentioned in YS III.13, then the present aphorism should be about "transformation of temporal dimensions". In that case the aphorism can be explained as follows.

That the mind could have anything as its object was its past dimension and that it has single-pointedness is its present aspect. The same mind has these two temporal aspects.

पुनः शान्तोदितौ तुल्यप्रत्ययौ चित्तस्यैकाग्रतापरिणामः ।।12।।

Punaḥ śāntoditau tulyapratyayau cittasyaikāgratāpariṇāmḥ.

Furthermore, the one-pointedness type of transformation of mind means the two similar experiences: the experience which has ceased and that which has arisen.

Comments

Here Patañjali is suggesting that the same phenomenon of entering into meditative trance can be described in terms of single-pointedness. But then, what is the single-pointedness of mind? Patañjali states it as this. There is a flow of similar cognitions. Hence the cognition that has ended (*śānta*) and the other that has arisen (*udita*) are similar (*tulya*). The two cognitions are similar because they have the same object.[9] Through this, Patañjali wants to suggest that when the mind passes from the worldly stage to meditative trance, the state (*avasthā*) of mind changes from many-pointedness to single-pointedness. But in the passage of single-pointedness, the mental state remains the same because the cognitions which arise are similar. The notion of "transformation of state" (*avasthāpariṇāma*) can perhaps be explained in this way.

In all the three aphorisms (YS II.9, 11 and 12), Patañjali is deliberately using the word *pariṇāma* (impact or effect, which is a transformation) in order to mark the difference from the Buddhist position (which talks about cessation or destruction; *nirodha* or *vināśa*) and underline the Sāṅkhya position. He wants to claim that in all these "changes" that take place in mind, mind does not cease, it only undergoes transformation. That is, mind-stuff—(the *sattva*-dominated manifestation of *prakṛti*)—which is *dharmī* (property-bearer) remains; only properties change. This has reference to Vasubandhu's discussion of the theme of "change" in AKB V.25–26. The theme will continue in the next aphorism where Vyāsa in his commentary will discuss the issue in detail.

एतेन भूतेन्द्रियेषु धर्मलक्षणावस्थापरिणामाः व्याख्याताः ।।13।।

Etena bhūtendriyeṣu dharmalakṣaṇāvasthāpariṇāmāḥ vyākhyātāḥ.

By this are explained the changes of the property (*dharma*), (temporal) characteristic (*lakṣaṇa*) and state of existence (*avasthā*) in the elements and sense-organs.

Comments

In the three previous aphorisms (YS III.9, 11, 12), Patañjali stated how one can explain the transformation that takes place in mind, when from an extravert state, one enters into meditative trance.

In this aphorism, he is suggesting an explanation of this transformation in terms of change of property (*dharmapariṇāma*), change of temporal characteristics (*lakṣaṇapariṇāma*)[10] and change of the state of existence (*avasthāpariṇāma*). Patañjali is saying here that transformations of material elements and also those of sense-organs can be explained on the same lines.

Patañjali's aphorism as well as Vyāsa's commentary clearly indicate that all this is a response to Vasubandhu's discussion of Sarvāstivāda in AK and AKB V.25–26.[11] *Sarvāstivāda* means the one who holds that objects belonging to all the three temporal dimensions (past, present and future) exist. Vasubandhu here presents four alternative explanations of the existence of a thing over the three times.

(1) *Bhāvānyathika*—only structure (*bhāva*) changes; the substantial nature (*drvyabhāva*) remains the same. This was the position of Bhadanta-Dharmatrāta. This corresponds to what Patañjali calls "*dharmapariṇāma*".
(2) *Lakṣaṇānyathika*—a thing has all the three (temporal) dimensions. When a thing possesses any one dimension, the other dimensions are not lost. They only remain latent. This corresponds to what Patañjali calls "*lakṣaṇapariṇāma*". This was the position of Bhadanta-Ghoṣaka.
(3) *Avasthānyathika*—temporal dimension of a thing is determined by the state of its existence, that is, its activity or functionality. (Here *avasthā* means activity, *kāritra*.) This was the position of Bhadanta Vasumitra. This explanation corresponds to what Patañjali calls "*avasthāpariṇāma*".
(4) *Anyathānyathika*—a temporal dimension of a thing is relative to temporal dimensions of other things. This relativist position belonged to Bhadanta Buddhadeva. Patañjali does not refer to this position.

Out of the four positions, Vasubandhu accepts the third position (*avasthānyathika*—explanation in terms of *kāritra*). He rejects other positions. He rejects the first position because it is nothing but *pariṇāmavāda* of Sāṅkhya.

Patañjali's position and also the position of the commentator Vyāsa seems to be that all the first three positions are defensible. Patañjali does not defend the relativistic (*anyathānyathika*) position. And though Vyāsa mentions the metaphors used for explaining the relativistic position,[12] he is not particular about defending it. But he tries to refute Vasubandhu's criticism of others.

(1) Vyāsa, while defending the first position (that is, which Vasubandhu calls the Sāṅkhya position), claims that *dharmī* exists over and above *dharmas*. There is no absolute eternality; there is no absolute destruction.
(2) Vyāsa, while defending the second position (*lakṣaṇānyathika* position), tries to refute Vasubandhu's charge that this position leads to admixture of temporal dimensions (*adhva-saṅkara*).[13] Vyāsa argues that the passions like love and anger can be simultaneously present; that does not mean that they should be simultaneously manifest. Hence, one temporal dimension may be manifest while others are simultaneously present, but non-manifest. Therefore, the admixture of temporal dimensions (*adhvasaṅkara*) is not a fault here.
(3) Vasubandhu says that the third alternative (*avasthānyathika*) is acceptable. But how to interpret it is a problem. Vasubandhu raises the following

problem: if past, present or future dimensions of a thing are determined by its activity (*kāritra*), how are past, present and future dimensions determined for an activity? By the activity of the activity? If the answer is, "The temporal dimensions are not to be applied to the activity", then the activity will be unconditioned (*asaṁskṛta*) and hence eternal (*nitya*).[14] Hence, the doctrine of activity itself comes to danger. Vasubandhu's answer to this problem is that "activity" and being of a thing (*dharma*) are not different. He does not accept *dharmin* apart from *dharma*. This is a Sautrāntika position. This answer is not acceptable to Vyāsa. Vyāsa tries to answer Vasubandhu's Sautrāntika view with the help of the transformationism (*pariṇāmavāda*) of Sāṅkhya. According to this theory, all change is transformation. When we say that a thing has changed, in fact only the form has changed, the basic thing has remained unchanged. Vyāsa gives the reason: "The quality-bearer can be eternal, but the qualities can be destructible and diverse".[15]

शान्तोदिताव्यपदेश्यधर्मानुपाती धर्मी ।।14।।

Śāntoditāvyapadeśyadharmānupātī dharmī.

The property-bearer is the one which runs through ceased, arisen and the properties which cannot be designated (as either ceased or arisen).

Comments

Patañjali, through this aphorism, is suggesting that the property-bearer can continue to exist even if its properties may arise or cease or may be in an indeterminate position. This is probably Patañjali's response to Vasubandhu's criticism of Pariṇāmavāda of the Sāṅkhyas.[16]

Vyāsa in his commentary explains the past, present and future properties of a property-bearer in terms of the activity (*vyāpāra*) of a thing:

(1) Present property is that which performs its activity.
(2) "The property-bearer has past properties", means some of its properties have already performed their activities and stopped (*kṛtvā vyāpāran uparatāḥ*).
(3) "Property-bearer has future properties"—means some properties which have arisen along with their activities immediately precede the future characteristics (*savyāpārā uditās te ca anāgatasya lakṣaṇasya samanantarāḥ*).

While explaining *avyapadeśya dharma*, Vyāsa asserts the Sāṅkhya thesis that "everything is everything". (Everything has potential to become everything else. Everything has properties of everything else in non-manifest form.) This meaning may not be intended by Patañjali.

Vyāsa here criticizes the Sautrāntika thesis that there are only properties without property bearer. According to him if there are only properties without interrelations among them, then there will be no experience of pleasure or pain (caused by past actions).[17]

TABLE 3.1 Three temporal paths according to Vyāsa and Vasubandhu

	Yogabhāṣya	*Abhidharmakośabhāṣya*
Past	*śāntāḥ ye kṛtvā vyāpārān uparatāḥ*	*yadā (kāritraṁ) kṛtvā niruddhās tadā'tītāḥ*
Present	*vartamānaḥ svavyāpāram anubhavan dharmī*	*Yadā (kāritraṁ) karoti tadā pratyutpannaḥ*
Future	*savyāpārā uditās te ca anāgatalakṣaṇasya samanantarāḥ*	*Yadā kāritraṁ na karoti tadā anāgataḥ*

Sources: YB III.14, AK and AKB V.26

Note: Vasubandhu says: "*adhvānaḥ kāritreṇa vyavasthitāḥ*" (AK V.26) "*yadā sa dharmaḥ kāritraṁ na karoti, tadā'nāgataḥ, yadā karoti tadā pratyutpannaḥ, yadā kṛtvā niruddhas tadā'tītaḥ*" (AKB on AK V.26) ("The temporal paths are explained by the notion of activity. When the particular thing does not perform the activity, it is future; when it performs, it is present; when it gets destroyed after performing it, it is past.)

However, Vyāsa's explanation of past, present and future is influenced by Vasubandhu's explanation. Vasubandhu explains these ideas in terms of *kāritra* (function, activity); Vyāsa talks in terms of *vyāpāra*. The similarity of the two explanations is shown in Table 3.1.

क्रमान्यत्वं परिणामान्यत्वे हेतुः ।।15।।

Kramānyatvaṃ pariṇāmānyatve hetuḥ.

Difference in the order (of properties) is the cause of the difference in the transformation.

Comments

Here Patañjali is differentiating between the three types of transformation he has already talked about, namely *dharmapariṇāma*, *lakṣaṇapariṇāma* and *avasthāpariṇāma*. Patañjali is suggesting here that when a thing changes, any of the three (or all the three) transformations take place. The properties may change, the temporal dimension may change or the state of existence may change. It is important to understand the distinction between them. Patañjali claims that each type of transformation represents a different sequential order. The point is not clear. Vyāsa explains:

(1) *Dharmapariṇāma*—properties change when for instance soil is transformed into a lump of clay, which in its turn transforms into an earthen pot which finally gets broken into pieces.
(2) *Lakṣaṇapariṇāma*—temporal dimensions change when for instance a future pot becomes present and the present lump of clay which is present, becomes past.
(3) *Avasthāpariṇāma*—the same thing gradually undergoes partial change and ultimately changes completely.

It may be noted here that Vasubandhu had discussed these aspects of transformation in AKB V.26 and Vyāsa had used this discussion in his commentary on YS III.13.

It can be said on the basis of the earlier discussion of the theme that Patañjali and, following him, Vyāsa have tried to synthesize the three different explanations of change given by Sarvāstivādins within the broad framework of Sāṅkhya.

Another point Vyāsa makes in his commentary is that there are two types of *cittadharmas* (characteristics of mind): directly known (*dṛṣṭa, paridṛṣṭa*) and not directly known (*adṛṣṭa, aparidṛṣṭa*). The directly known characteristic of mind is cognition/consciousness (*darśana*). Others can *be* called ontological characteristics (*vastumātrātmaka*). There are seven according to him:

(1) *Nirodha*: mind is subject to cessation.
(2) *Dharma*: it is a phenomenon.[18]
(3) *Saṁskāra*: it is conditioned.
(4) *Pariṇāma*: it undergoes transformation.
(5) *Jīvana*: it contains life force.
(6) *Ceṣṭā*: it stimulates behavior.
(7) *Śakti*: mind has capacity (to concentrate/meditate).

From this list, the first three seem to be obviously influenced by Buddhism. The fourth characteristic, *pariṇāma*, is also found in the Buddhist literature but originally it seems to be of Sāṅkhya type. *Jīvana* reflects *jīvitendriya* which is a common mental factor (*sādhāraṇa cetasika*) according to the Buddhist Abhidharma. *Ceṣṭā* is physical action, which is an expression of *cetanā* (volition). And *śakti* would be the capacity for *ekāgratā*. Both *cetanā* and *ekāgratā* are common mental factors according to Abhidharma Buddhism.

परिणामत्रयसंयमादतीतानागतज्ञानम् ।।16।।

Pariṇāmatrayasaṅyamād atītānāgatajñānam.

By integrated concentration upon the three-fold transformations comes the knowledge of past and future.

Comments

The present is causally related to the past and future due to which the past transforms into the present and the present into the future. *Dharmapariṇāma, lakṣaṇapariṇāma* and *avasthāpariṇāma* are the three ways of understanding this transformation. Patañjali claims that by applying triple concentration to the three forms of transformation one can know past and future.

In *Abhidharmakośabhāṣya*, the issue is raised about Buddha's knowledge about future happenings in one's life.[19] Here Vasubandhu talks about three ways:

(1) inference based on the causal connection between the past and the present,[20] (2) revelation of future based on the present signs and (3) direct, unconditioned revelation.

(1) Inference: Vasubandhu explains the futuristic inference as follows. The Buddha knows:

> In the past this type of action was performed; from that such and such type of maturation has taken place. Or from this type of (past) *dharma*,[21] that type of *dharma* is taking place. And presently there is such and such type of action; so, such and such type of maturation will take place in future. Or from this type of (present) *dharma*, this type of *dharma* will take place in future.

The problem with this position according to Vasubandhu is that then the Buddha's futuristic knowledge will be inferential.

(2) The second alternative is that the Buddha knows future by revelation, when he sees certain signs in the present. The objection against this view would be that the Buddha would not know the final end (*aparānta*) without knowing the beginning (*pūrvānta*).[22] He will be a "knower through signs" (*naimittika*) and not the intuitive-seer (*sākṣātkārī*).

(3) Vasubandhu then gives the final view from the Sautrāntika viewpoint that the Buddha knows everything merely by his wish.[23]

In this discussion Vasubandhu mentions knowledge through meditation or supernormal cognition as a possibility (when he says that the Buddha's knowledge of future supersedes supernormal cognition), but does not elaborate on that.

Patañjali and Vyāsa seem to be giving their explanation as to how the knowledge of past and future is possible through supernormal type of knowledge, which Buddhists would include in *abhijñā*. However, their explanation is based on causal connections between past and present and those between present and future. So the question remains as to how it is different from inferential knowledge.

शब्दार्थप्रत्ययानामितरेतराध्यासात्संकरस्तत्प्रविभागसंयमात् सर्वभूतरुतज्ञानम् ||17||

Śabdārthapratyayānām itaretarādhyāsāt saṅkaras tatpravibhāgasaṃyamāt sarvabhūtarutajñānam.

There is admixture of word, object (meaning) and the cognition thereof, because of superimposition on each other. If integrated concentration is practiced on each separately, one has the knowledge of (the meanings of) the utterances of all beings.

Comments

Here Patañjali is advocating a view of language which is very close to Buddhism. Realist Buddhists (Vaibhāṣikas and Sautrāntikas) accepted external objects (*artha*). But according to them words do not refer to real objects. According to Sautrāntikas the real objects are unique particulars whereas the words express the unreal universals, which are the mental constructs (*vikalpa*). Hence, words and objects are essentially different and unrelated. Similarly, *pratyaya* (cognition) of an object is different from both the external objects and words. But it is our general tendency to mix up the three. Many pseudo-objects appear to be real because they are expressed by words. Buddhists call them *prajñapti-sat*. In fact, the whole practice of language takes place by mixing the three elements.[24] It follows that in order to understand how language functions, we should separate the three elements from each other and see how they operate. All this seems to be clear in the Buddhist framework. Though we find this clearly in later Buddhist logicians like Diṅnāga and Dharmakīrti, its roots are traceable in the writings of Asaṅga and Vasubandhu.

Patañjali seems to be claiming that if we apply integrated concentration to the distinct factors (namely words, objects and cognitions) which get mixed up in linguistic practice, then we can develop an insight into the functioning of language and, as a result, understand the languages of all living beings. Though the claim is not convincing, it is consistent with Patañjali's general way of theorizing about supernormal powers.

संस्कारसाक्षात्करणात्पूर्वजातिज्ञानम् ।।18।।

Saṃskārasākṣātkaraṇāt pūrvajātijñānam.

By realization of impressions arises the knowledge of previous births.

Comments

The word *saṁskāra* is used in two contexts. (1) In the context of memory: a cognition generates an impression which remains dormant. When it is awakened by some stimulus, it generates recollection. (2) In the context of fruition of actions: an action, whether good or bad, causes latent impressions, which, when mature, cause pleasure and pain. Since Patañjali is referring to the knowledge of the past, the first context seems to be more relevant. Here Patañjali seems to be reflecting the supernormal knowledge accepted in Buddhism called *pūrvenivāsānusmṛti* (the recollection of the previous births).

Asaṅga, in *Śrāvakabhūmi*,[25] describes the method of acquiring the knowledge of previous births. It involves remembering events of one's childhood in as elaborate detail as possible and then practicing meditative realization on them. Patañjali's notion of *saṁskārasākṣātkaraṇa* (realization of impressions) seems to refer to that.

Vyāsa interprets the term *saṁskāra* as *dharma* and *adharma* (merit and demerit), which refers to the second context, which does not seem to be directly relevant here.

प्रत्ययस्य परचित्तज्ञानम् ।।19।।

Pratyayasya paracittajñānam.

(By practicing integrated concentration) on experience arises the knowledge of other minds.

Comments

Paracittajñāna (knowledge of another mind) is one of the supernormal knowledges according to Buddhism. Patañjali says that it can be attained through the integrated concentration on experience/cognition (*pratyaya*).

Asaṅga in *Śrāvakabhūmi*[26] explains the procedure of acquiring the knowledge of other minds which involves a keen observation of the correlation between mental states and physical expressions, which is a more rational method.

Patañjali's cryptic reference to *pratyaya* could be suggestive of this procedure.

न च तत्सालम्बनं तस्याविषयीभूतत्वात् ।।20।।

Na ca tat sālambanaṃ tasyāviṣayībhūtatvāt.

But this (knowledge of other minds) does not pertain to the content (of another person's experience) because that is not the object (of his integrated concentration).

Here Patañjali means, as Vyāsa explains, that through the knowledge of another mind one understands the character of the other's mental state, not the object of it. For instance, one realizes that the other's mind is attached, not the object (*ālambana*) to which it is attached.

This clearly echoes Vasubandhu's explanation:

> Does the knowledge of another mind grasp the form or the object of the other mind? It being irrespective of form or object, knows that "the mind is attached", it does not know that it is attached to *this* physical form". Otherwise the mind will have physical form also as its object. And that which grasps the object of the other mind, will also grasp the nature of the thing.[27]

कायरूपसंयमात्तद्ग्राह्यशक्तिस्तम्भे चक्षुष्प्रकाशासम्प्रयोगेऽन्तर्धानम् ।।21।।

Kāyarūpasaṅyamāt tadgrāhyaśaktistambhe cakṣuṣprakāśāsamprayoge'ntardhānam.

By integrated concentration on the appearance of the body, when its power to be perceived is stopped, the body loses contact with light of the eyes and there is disappearance.

Comments

Patañjali here is describing "disappearance of body" as the supernormal achievement. Patañjali explains the procedure as having three steps: (1) triple concentration on the visible form of the body, (2) suspension of the visibility of the body and (3) the body getting unconnected with the light of the eyes.

Asaṅga explains the same supernormal achievement. "One passes through a wall or a building without his body sticking to them". He explains the process in terms of five-fold resolution or conviction: (1) ideation of lightness, (2) ideation of softness, (3) ideation of space, (4) ideation of identification of mind and body and (5) ideation of resolving one thing as another. Asaṅga explains that by the first four ideations one's body becomes lighter, softer, quicker, more glittering, associated with mind, more connected with mind, more regulated by mind.[28]

Hence, the two procedures are different. Asaṅga's is more elaborate. Patañjali's procedure seems to be more simplified, but more mystical.

सोपक्रमं निरुपक्रमं च कर्म तत्संयमादपरान्तज्ञानमरिष्टेभ्यो वा ।।22।।

sopakramaṁ nirupakramaṁ ca karma tatsaṅyamād aparāntajñānam ariṣṭebhyo vā.

An action either advances (to give fruit) or does not advance (to give fruit). By practicing integrated concentration on it knowledge of the latter, end (of life) arises; or (it arises) from the signs of approaching death (*ariṣṭa*).

Comments

Aparānta means final end. The words *pūrvānta* and *aparānta* are used in Buddhist literature in the sense of beginning and end (respectively) of any event or series of events or the course of transmigration (*saṁsāra*) as a whole.

Patañjali here is using the word *aparānta* in the sense of the end of life, that is, death. He is talking here of precognition of death as a kind of supernormal knowledge.

The precognition of death can be considered as a special case of precognition (the knowledge of future). Patañjali has already talked about "the knowledge of past and future" in YS III.16. We have seen there that the knowledge of past and future is possible because they are causally connected with the present. The same idea applies here.

Death is the end of a lifespan and lifespan is the outcome of karma (according to both Buddhism and Patañjali). Hence, Patañjali states that if one knows the working of karma in the case of a person, then one can have precognition of his death. The Buddhists, too, believed that such a precognition is possible and that the Buddha in particular did have it.[29]

Here Patañjali distinguishes between karma which has started giving fruition (*sopakrama*) and that which has not (*nirupakrama*). Then he introduces his usual technique of *saṁyama*.

The other basis of the pre-knowledge of death that Patañjali suggests is what is called *ariṣṭa* (the signs indicating forthcoming death). This is the theme related to medical science and not the theory of supernormal power. Various types of physiological signs indicating forthcoming death have been elaborately discussed in various chapters of the *indriyasthāna* section of the *Carakasaṁhitā*. Patañjali seems to be referring to that. Vyāsa in his commentary refers to *ariṣṭa*s in terms of the Sāṅkhya-style classification of them into *ādhyātmika* (self-related), *ādhibhautika* (related to other ordinary beings) and *ādhidaivika* (related to divine beings).

मैत्र्यादिषु बलानि ।।23।।

maitryādiṣu balāni.

(Different types of) strength are achieved through (integrated concentration) on friendliness etc.

Comments

By *maitryādi*, Patañjali seems to refer to the four sublime attitudes, namely *maitrī*, *karuṇā*, *muditā* and *upekṣā*, as indicated by Vyāsa. He has recommended them as the objects of meditative practice (*bhāvanā*) in YS I.33. Here he is saying that by applying triple concentration technique to them, one achieves special powers (*balāni*).

In Buddhism, the four sublime attitudes are called *apramāṇa* (immeasurable). *Apramāṇas* are included in Buddhism in the list of objects of meditation (*karmasthāna*) and some special importance is attached to them. Asaṅga in *Abhidharmasamuccaya* makes a list of special qualities (*vaiśeṣika*1-*guṇas*) in which the four immeasurable attitudes come first. About the four immeasurables, Asaṅga says that even a practitioner of a non-Buddhist sect can accomplish them by applying the four stages of pure meditation.[30]

Asaṅga also explains what one achieves through the four immeasurables:

> What does one perform through immeasurables? He abandons the opposite attitudes (such as attachment, greed hatred, and envy). By living with compassion, he fills up the stock of merit and does not get upset while observing the maturation of the karma of the beings.

Among the four immeasurables, *maitrī* has been given a special status in the Buddhist theory.

Tandon (1998: 15) refers to psychological and psychosomatic advantages of *maitrī* (*mettābhāvanā*; loving kindness) thus:

> One sleeps in comfort, wakes in comfort, dreams no evil dreams, is dear to human beings, is dear to non-human beings, is guarded by deities, remains un-affected by fire, poison and weapons and though one may penetrate not the beyond, one reaches the Brahmā World.

It is possible that by the word *balāni*, Patañjali means some other supernormal powers such as Buddha's powers, because in the next aphorism he talks about "elephant power" (*hastibala*), which is a parameter of Buddha's physical power.

Vyāsa in his commentary makes an interesting point which marks discrepancy between *Sūtra* and *Bhāṣya*. Vyāsa remarks that out of the four attitudes, namely *maitrī* (loving kindness towards the happy ones), *karuṇā* (compassion towards the unhappy ones), *muditā* (gladness towards the virtuous ones) and *upekṣā* (indifference towards the sinful ones), only the meditative cultivation (*bhāvanā*) of the first three attitudes can give rise to irresistible powers. But the last one, namely indifference towards the vicious ones, is just an attitude; it cannot be the matter of meditative cultivation and hence it cannot generate any supernormal power.[31]

The view of Vyāsa is inconsistent with Patañjali's main statement about the four attitudes in YS I.33 according to which all the four attitudes are the matter of meditative cultivation. The discrepancy arises because, as I have suggested while commenting on the said aphorism, Patañjali accepts universalist interpretation of the four attitudes whereas Vyāsa follows specificist interpretation.

बलेषु हस्तिबलादीनि ।।24।।

baleṣu hastibalādīni.

The strength of elephant etc. (can be acquired) (by practicing integrated concentration) on strengths.

Comments

Here Patañjali is talking about extraordinary physical strength as the supernormal power. His special mention of elephant power may have a Buddhist background.

In Buddhist literature *hastibala* (elephant-power) was used as a measurement for measuring supernormal physical strength. Buddha's physical power is called *nārāyaṇabala*. Vasubandhu calculates Buddha's power in terms of *hastibala* as follows:

hastibala × 10 = *gandhahastibala*[32]
hastibala × 20 = *mahānagnabala*[33]
hastibala × 30 = *praskandibala*[34]
hastibala × 40 = *varāṅgabala*[35]
hastibala × 50 = *cāṇūrabala*[36]
hastibala × 60 = *nārāyaṇabala* (=Buddha's power)[37]

When Patañjali says "*hastibala* etc.", he might be meaning thereby different degrees of power (which start with *hastibala*), which Vasubandhu talks about. Vyāsa imagines different examples, such as *vainateyabala* (eagle-power) and *vāyubala* (the power of air). The source is unknown.

प्रवृत्त्यालोकन्यासात्सूक्ष्मव्यवहितविप्रकृष्टज्ञानम् ।।25।।

pravṛtyālokanyāsāt sūkṣmavyavahitaviprakṛṣṭajñānaṃ.

Knowledge of the subtle, hidden and distant objects arises by focusing (mind) on the light generated by the activity (of visual sense organ).

Comments

According to Sāṅkhya, an existent object is not perceived under the following conditions:

(1) When the object is too distant
(2) When the object is too close
(3) When the sense-organ is damaged
(4) When the mind is not attentive
(5) When the object is too subtle
(6) When there is an obstacle between the sense-organ and the object
(7) When the object is influenced by another dominant object
(8) When similar objects are mixed together[38]

Here Patañjali is referring to three conditions: (1) when the object is subtle (*sūkṣma*), (2) when there is an obstacle (*vyavahita*) and (3) when the object is too distant (*viprakṛṣṭa*).

Buddhists had developed the notion of *divyacakṣu* (divine eye) which they applied mainly to the power to visualize the birth and death of beings remote from senses. Jainas had introduced the extrasensory perception (clairvoyance) which they called *avadhijñāna*. Patañjali claims that such a clairvoyant knowledge is possible by focusing (mind) on the light generated by the activity (of visual sense organ).

भुवनज्ञानं सूर्ये संयमात् ।।26।।

bhuvajñānaṃ sūrye saṁayamāt.

The knowledge of the cosmos arises by practicing integrated concentration on the Sun.

Comments

In this aphorism and many aphorisms to follow, Patañjali is following, for introducing different supernormal powers, a pattern of thinking which can be called "single key-element". The pattern can be described as follows. Given any system, we can identify an element which functions like a key to understand the whole system; or which appears to control the system. Patañjali seems to believe that by applying the technique of triple concentration to the key-element of a system, a yogin can know the whole system or control the whole system.[39] How Patañjali follows this pattern is shown in Table 3.2.

TABLE 3.2 The correlation between the objects of meditation and supernormal powers according to Patañjali

Sūtra no.	*System to be known or controlled*	*Key element*	*The possible justification*
III.26	*Bhuvana* (cosmos)	Sun	The sun appears to illuminate the cosmos.
III.27	The gathering of stars	Moon	The moon appears as the largest shining heavenly body at night accompanied by other stars.
III.28	The motions of stars	Dhruva (pole)	The pole-star does not move when other stars move.
III.29	The bodily system	Naval circle	The naval is the location of the umbilical cord through which the fetus gets nourishment.
III.30	Hunger and thirst	The throat–pit	Throat is the passage of eaten food and drunk water.
III.31	Steadiness of body	Tortoise channel	Tortoise body has steadiness.
III.32	Realized beings (*siddhas*) all over the universe	The light in the top of the head	Spiritual excellence is cognized at the top of the head and realized beings represent spiritual excellence.

Sources: *Yogasūtra* aphorisms mentioned in the first column

The fourth column gives possible justifications. (They are not supposed to be scientific justifications. Some of them appear to be based on imagination.)

Vyāsa's commentary on this aphorism is unnecessarily elaborate. Here Vyāsa has given an exposition of his "*bhuvana-jñāna*" (his knowledge of the cosmos). It is largely mythological, and partly based on the geographical knowledge which was available to him. The point that is important for our purpose is that we also have Buddhist cosmology reflected in Vasubandhu's *Abhidharmokoṣabhāṣya* Chapter III. It is also an admixture of mythology and geography. Vyāsa's picture of the cosmos is partly governed by Vedic-Purāṇic cosmology and partly by the Buddhist cosmology. It is also possible that Vasubandhu's cosmology was partly influenced by Brahmanical cosmology. It is worthwhile to compare Vyāsa's picture of the cosmos with that of Vasubandhu. The comparison can be divided into three parts: the lower worlds, the middle world and the higher worlds. Tables 3.3, 3.4 and 3.5 give a picture of the three worlds respectively. Similar terms used by Vyāsa and Vasubandhu are given in bold letters in these tables.

Hierarchy of *svargas*

Both Vasubandhu and Vyāsa talk about the hierarchy of *svargas*, though the names are different. In Vasubandhu's scheme the lowest *svarga* belongs to *Kāmadhātu* (the world of sensuous pleasure); the higher one belongs to *Rūpadhātu* (the world of

TABLE 3.3 Lower worlds according to Vyāsa and Vasubandhu

	Vyāsa	*Vasubandhu*
The lowest world	***Avīci* (-Naraka)**	***Avīci* (-Naraka)**
Above *Avīci*	Six ***Mahānarakas*** and seven *pātālas*	Seven ***Mahānarakas*** (no mention of *pātālas*) but Vasubandhu refers to *Pretaloka* and *Yamaloka*
Names of *Narakas* above *Avīci*	(From top to bottom:) *Mahākāla*, *Ambarīṣa*, ***Raurava***, ***Mahāraurava*** ***Kālasūtra***, Andhatāmisra	(From top to bottom:) *Saṁjīva*, ***Kālasūtra***, *Saṅghātra*, ***Raurava***, ***Mahāraurava***, *Tapana*, *Pratāpana*

Sources: YB III.26, AKB III.1

TABLE 3.4 Middle world (between Naraka and *svarga*)

	Vyāsa	*Vasubandhu*
Mountain in the middle	***Sumeru***	***Sumeru***
The island (*dvīpa*) adjacent to Sumeru	***Jambūdvīpa***	***Jambūdvīpa***
Other *dvīpas*/*varṣas*	Three lands in the intervals between the northern mountains—*Ramaṇakaṁ*, *Hiraṇmayaṁ*, ***Uttara-kurus***	*Pūrvavideha*, *Avara-godānīya*, ***Uttarakuru***
Waters covering Jambūdvīpa	That Jambūdvīpa is covered by **a double-sized sea of salty water**	Outside it is **the great sea of salty water**

Sources: YB III.26, AKB III.1, 52d

TABLE 3.5 Description of the *svarga* on Sumeru

Vasubandhu	*Vyāsa*
In the midst of the plane of ***Sumeru*** there is a city called ***Sudarśana***. It is the capital of Śakra, the king of gods. The palace of Śakra, the king of gods, is called ***Vaijayanta***. There are four gardens, which are the playgrounds of gods. They are named as ***Citraratha***, *Pāruṣyaka*, ***Miśrakāvana*** and ***Nandana-vana***. The assembly of gods is in the south-west called ***Sudharmā***.	***Sumeru*** is the garden-site of gods. There are the following gardens: ***Miśravana***, ***Nandana***, ***Caitraratha*** and *Sumānasa*. The assembly of gods is called ***Sudharmā***. The city is called ***Sudarśana*** and the palace is called ***Vaijayanta***.

Sources: AK and AKB III.66–68, YB III.26

TABLE 3.6 The lowest *svarga* as the *svarga* of enjoyment and the kinds of gods who reside in it

Vasubandhu	*Vyāsa*
Gods residing in the lower *svarga* (*Kāmadhātu*): *Catur-mahārāja-kāyika*, ***Trāyastriṁśa***, ***Yāma***,***Tuṣita***, ***Nirmāṇarati*** and ***Paranirmitavaśavartin*** **(six in number)**. They are said to enjoy pleasures (***kāmān paribhuñjate***) by way of making pairs, embracement, joining hands, smiling, gazing, and copulating.	**Gods residing in the lower *svarga* (*Māhendra*):** ***Tridaśā***, *Agniṣvātta*, ***Yāma***, ***Tuṣta***, ***Aparinirmitavaśavartin***, and ***Parinirmitavaśavartin*** **(six in number)**. They get what they want (*saṅkalpasiddha*). They enjoy pleasures (***kāmabhoginaḥ***). The best and favourable nymphs are at their service.

Sources: AK and AKB III.69, YB III.26

Notes:

(1) *Nirmāṇarati* (those who enjoy their own magical creations) and *paranirmitavaśavartin* (those who control the magical creations of others) are the two categories of gods belonging to *Kāmāvacaraloka* (lower *svarga*) according to Buddhism. The terms "*aparinirmitavaśavartin*" and "*parinirmitavaśavartin*", that Vyāsa uses seem to have been derived from the Buddhist terms.

(2) *Trāyastriṁśa* of Buddhism (which means thirty-three) and *Tridaśa* (which can be interpreted as thirty) of the Vedic tradition might have some relation or parallelism.

forms) and the highest one to *Ārūpyadhātu* (the world of formlessness). Since this categorization of the three worlds is not acceptable to Vyāsa, he replaces it with a categorization based on Brahmanical cosmology. Accordingly, *Māhendra* is the lowest *svarga*. Above it is *Prājāpatya*. And the highest is *Brāhma* which consists of Jana, *tapas* and Satya.

One common point between the two schemes is that the lowest *svarga* in both is the world of enjoyment, whereas the higher *svarga*s are characterized by meditative experience. Table 3.6 indicates the similarity between the lower *svarga*s as described by Vasubandhu and Vyāsa.

Higher *svargas* are related to meditative stages

Vyāsa arranges the higher *svarga*s differently from Vasubandhu. But there are important points of contact.

One important point of contact is that in Vasubandhu's hierarchy of higher *svarga*s, there are four levels related to what Buddhists would call *rūpadhyāna*. Vyāsa conceives of the four levels as marked by the gods who survive by meditation as the food (*dhyānāhārāḥ*). There is similarity between the titles of gods residing in them as shown by Table 3.7.

One difference between the two accounts of higher *svarga*s is that Vasubandhu's four types of higher worlds have direct correlation with the four stages of form-meditation (*rūpadhyāna*) and the highest world correlates with the formless meditation according to Buddhism.

TABLE 3.7 Gods residing in higher *svargas* according to Vasubandhu and Vyāsa

Higher world no.	*Abodes* (sthāna/bhūmi) *of gods in the respective* svarga *according to Vasubandhu (AKB III.2)*	*Abodes of gods* (devanikāya) *in the respective* svarga *according to Vyāsa*
O Mahat-loka	--------	*Kumuda, Ṛbhu, Pratardana, Añjanābha, Pracittābha*
1	***Brahmakāyika*, *Brahmapurohita*,** *Mahābrahmā*	***Brahmapurohita*, *Brahmakāyika*,** *Ajara, Amara*
2	*Parīttābha, Apramāṇābha,* ***Ābhāsvara***	***Ābhāsvara***, *Mahābhāsvara, Satyamahābhāsvara*
3	*Parīttaśubha, Apramāṇa-śubha, Śubhakṛtsna*	*Acyuta, Śuddhanivāsa, Satyābha Saṁjñāsaṁjñin*
4	*Anabhraka, Puṇyaprasava, Bṛhatphala, Abṛha, Atapa, Sudṛśa, Sudarśana, Akaniṣṭha*	No mention of the fourth world
5	*Ārūpya* (no abode)	*Videha* and *Prakṛtilaya* (belong to the realm of liberation and hence not to be counted as world)

Sources: AKB III.2, 3a; YB III.26

TABLE 3.8 Abodes of gods belonging to the third (highest?) *svarga* and the objects of their meditational enjoyment according to Vyāsa's account

Abode of gods	*Object of meditation*
Acyuta	*Vitarka*
Śuddhanivāsa	*Vicāra*
Satyābha	*Ānanda*
Saṁjñāsaṁjñin	*Asmitā*

Source: YB III.26

Vyāsa on the contrary correlates different stages of *dhyāna* with the third higher world only. Vasubandhu's account is more systematic whereas there is a discrepancy in Vyāsa's account. That is because, on the one hand, Vyāsa says that all the stages of higher *svargas* are marked by gods who survive on meditation (*dhyānāhāra*). But he correlates the four types of *dhyāna* only with the third stage of higher *svarga*. This is indicated by Table 3.8.

So, it remains unclear as to what kind of *dhyāna* is practiced by the gods belonging to first two stages or even to the 0th stage (*Mahat-loka*).

Observations

Vyāsa might have presented the picture of the cosmos, partly on the basis of Vedic-Purāṇic tradition, but his picture of the cosmos was also influenced by the Buddhist picture as elaborated by Vasubandhu in AKB III. The main doctrine

Vyāsa seems to borrow from Vasubandhu is the ideas of lower *svarga* dominated by sensuous enjoyment and higher *svarga* dominated by meditative trance. Buddhists divided the universe in three *dhātus*, namely *kāmadhātu*, *rūpadhātu and ārūpyadhātu*; so that the lowest *svarga* is the highest stage in *kāmadhātu*, whereas the higher *svargas* belong to *rūpadhātu* and *arūpadhātu*. Vyāsa avoids the terminology of three *dhātus* and uses different terminology. But the terminology of *kāmabhuj* in the description of the gods belonging to the lower *svarga* and the terminology of *dhyānāhāra* (consuming meditation as food) in the description of the higher *svargas* clearly indicate the influence of Buddhism. Vyāsa tries to explain the beings belonging to higher *svargas* in Sāṅkhya terms (*bhūtavaśinaḥ*, *bhūtendriyavaśinaḥ*, *bhūtendriyaprakṛtivaśinaḥ* etc.), but he also uses the meditational terminology of *vitarka*, *vicāra*, *ānanda* and *asmitā* which has affinity with Buddhism. The Buddhist description of the higher *svargas* seems to be more systematic and original.

चन्द्रे ताराव्यूहज्ञानम् ।।27।।

Candre tārāyūhajñānam.

(As a result of integrated concentration) on the moon, arises the knowledge of the arrangement of stars.

Comment

Here Patañjali is treating the moon as the key object for having insight into the arrangement of heavenly bodies. (See the first part of my comment on the aphorism YS III.26 indicating what kind of reasoning seems to underlie Patañjali's claims of this kind. The same principle operates till YS III.31.) Treating moon as the key-object seems to be based on the popular belief that moon is the leader of heavenly bodies (*tārānātha*) seen at night.

ध्रुवे तद्गतिज्ञानम् ।।28।।

Dhruve tadgatijñānam.

(As a result of the integrated concentration) on the pole-star, arises the knowledge of their movements.

Comment

Here Patañjali is treating the pole-star as the key-object for having an insight into the movements of heavenly bodies. Treating pole-star as the key-object seems to be based on the belief that since the pole-star is firm when all other heavenly bodies move, the former must be controlling the movements of the latter.

नाभिचक्रे कायव्यूहज्ञानम् ।।29।।

Nābhicakre kāyavyūhajñānam.

(As a result of the integrated concentration) on the navel circle, arises the knowledge of the structure of the body.

Comment

Here Patañjali is treating the naval circle as a key-object for understanding the structure of the body. Treating a naval circle as a key-object seems to be based on the role of the umbilical cord connected with it though which a fetus gets nourishment or on the idea that the digestive system in the stomach controls the bodily functions and the naval circle is at the centre of the stomach.

कण्ठकूपे क्षुत्पिपासानिवृत्तिः ।।30।।

Kaṇṭhakūpe kṣutpipāsānivṛttiḥ.

(As a result of the integrated concentration) on the gullet (literally: pit of the throat) arises the sub-dual of hunger and thirst.

Comment

Here Patañjali is claiming that integrated concentration on the throat pit leads to the control over hunger and thirst. Treating the throat as a key-object seems to be based on our experience of hunger and thirst in that location of the body.

कूर्मनाड्यां स्थैर्यम् ।।31।।

Kūrmanāḍyāṃ sthairyam.

(As a result of the integrated concentration) on the bronchial tube arises steadiness.

Comments

According to Vyāsa, *kūrmanāḍī* refers to the channel in the chest below the pit of the throat. It is called *kūrmanāḍī* because it has a tortoise shape. The significance of the name of the channel could also be related the slow motion or steadiness of tortoise. Some modern interpreters have associated the channel with continuous concentration on breathing without applying any force, due to which breathing considerably slows down resulting into stillness of mind and body. This interpretation seems to be influenced by *ānāpāna* meditation of Buddhism. (See the comments on YS II.50.)

मूर्धज्योतिषि सिद्धदर्शनम् ।।32।।

Mūrdhajyotiṣi siddhadarśanam.

(As a result of integrated concentration) on the light in the head, one has the vision of perfect beings.

Comments

Patañjali's claim about integrated concentration on the light in the head seems to follow the same principle as the preceding aphorism. Here the reasoning underlying Patañjali's claim seems to be the idea that spiritual excellence is realized by yogins at the top of the head and that the perfect beings (*siddhas*) represent spiritual excellence.

प्रातिभाद्वा सर्वम् ।।33।।

Prātibhād vā sarvam.

Or as a result of intuitive knowledge, (the yogin knows) all.

Comments

Patañjali claims that a yogin can know everything through *prātibhajñāna*. Here Patañjali could be referring to what Vasubandhu refers to as *pratibhānapratisaṁvit* (intuitive accurate knowledge) in AKB. After discussing eighteen distinctive characteristics (*āveṇikadharmas*) of the Buddha, Vasubandhu describes Buddha's other qualities which he shares with his disciples.[40] He discusses there six qualities, all related to special capacity for knowledge.

(1) *Araṇā*: it is the knowledge of previous births and the cycle of births and rebirths (*saṁvṛtijñāna*) useful for removing strife (*raṇa*) of others.
(2) *Praṇidhijñāna*: knowledge of anything that the meditator gets by making a resolve "I want to know this"[41] and then by attaining the ultimate (*prāntakoṭikam*) fourth meditative trance.
(3) *Dharmapratisaṁvid* (accurate knowledge of phenomena).
(4) *Arthapratisaṁvid* (accurate knowledge of meanings).
(5) *Niruktipratisaṁvid* (accurate knowledge of linguistic expressions).
(6) *Pratibhānapratisaṁvid* (intuitive accurate knowledge).

The last one is *Pratibhānapratisaṁvid*. About it Vasubandhu says that out of ten types of knowledge, it consists of nine types (*nava jñānāni*) (except the knowledge of cessation)[42] and also that it has all of the planes of existence as its scope.[43] Vasubandhu also says that for developing intuitive knowledge, the study of *hetuvidyā* (logic) is a prerequisite.[44]

This seems to be at the background when Patañjali says that everything can be known with the help of *prātibhajñāna*.

हृदये चित्तसंवित् ।।34।।

Hṛdaye cittasaṁvit.

(By practicing integrated concentration) on the heart, arises awareness of the mind.

Comments

It seems to be a part of common belief, and also the Buddhists' belief, that the heart is the location of mind. According to Rhys Davids and Stede the word *hadaya* (Pali analogue of the Sanskrit *hṛdaya*) refers to "the heart as seat of thought and feeling, especially of strong emotion (as in Vedas!), which shows itself in the action of the heart". In Pali literature, the word *hadaya* is sometimes used as a synonym for mind (*citta*) ("*hadayaṃ vuccati cittam*"[45]).

सत्त्वपुरुषयोरत्यन्तासंकीर्णयोः प्रत्ययाविशेषो भोगः परार्थत्वात्स्वार्थसंयमात् पुरुषज्ञानम् ।।35।।

Sattvapuruṣayor atyantāsaṃkīrṇayoḥ pratyayāviśeṣo bhogaḥ parārthatvāt svārthasaṁyamāt puruṣajñānam.

Sensuous experience consists in the common experience shared by mind (*sattva*) and spirit (*puruṣa*) (which are in fact) absolutely distinct from each other; and because it (that is, mind) exists for someone else (that is, for the sake of *puruṣa*), by practicing integrated concentration on "for itself", one knows the spirit (*puruṣa*).

Comments

This supernormal achievement presupposes Sāṅkhya metaphysical framework. In Sāṅkhya the basic distinction is between *prakṛti* (the insentient matter) and *puruṣa* (spirit, which is of the nature of pure witness consciousness). In Yoga, this metaphysical distinction assumes a more phenomenological form as the distinction between "*buddhi* and *puruṣa*" or "*sattva* and *puruṣa*" (*buddhi* is sometimes called *sattva*, because it is dominated by *sattvaguṇa*, which represents light and cognition). *Buddhi* or *sattva* in Sāṅkhya is a cognitive manifestation or cognitive aspect of *prakṛti*. Among difference aspects of *prakṛti*, the cognition aspect is closest (in terms of similarity) to *puruṣa*. Hence, knowing the distinction between them is crucial for developing emancipating insight. Here by saying that *sattva* and *puruṣa* are absolutely distinct (*asaṅkīrṇa* = non-mixed-up) Patañjali is emphasizing the same distinction. In the stage of sensuous experience (of enjoyment or suffering: *bhoga*) they get mixed up—and yield a common experience (*pratyaya-aviśeṣa*). The term *parārtha* (which means, "for the sake of the other") has special significance in Sāṅkhya. Accordingly, *prakṛti* produces its manifestations for the sake of the other (i.e. for *puruṣa*). By contrast, *puruṣa* does not exist for the sake of the other but for itself. Though *prakṛti* is for the sake of the other (*parārtha*), *puruṣa* shines for itself (*svārtha*). Hence integrated concentration on this notion of "for itself" gives an insight into the nature of *puruṣa*. This is how the insight into *puruṣa* becomes possible, according to Patañjali. This insight into *puruṣa* yields different supernormal powers. This is what Patañjali indicates in the next aphorism.

ततः प्रातिभश्रावणवेदनादर्शास्वादवार्ता जायन्ते ।।36।।

Tataḥ prātibhaśrāvaṇavedanādarśāsvādavārtā jāyante.

Thence are produced intuitive knowledge, (supernormal) auditory knowledge, sensation, sight, taste and information.

Comments

Here Patañjali is referring to some of the powers which he has already spoken about before or he is going to speak about later. For example, he has spoken about intuitive knowledge (*prātibha*);[46] he will talk about supernormal auditory cognition (*śrāvaṇa*) ("Divine ear", YS III.40). He has talked about sensation of the whole body (*vedana*) (YS III.29) and also about the perception of remote objects (*ādarśa*) (YS III.24). Here he is adding two powers to the list: special taste (*āsvāda*)[47] and information about remote happenings (*vārtā*). Patañjali is claiming that all these powers arise from the knowledge of the spirit (*puruṣa*).

I have accepted the meaning of the term *vārtā*, which seems most natural. In the Vyāsa's commentary, however, there is confusion on this. According to the commonly accepted version, Vyāsa interprets *vārtā* as "divine odour" (and he interprets *vedana* as divine touch) and, in this way, tries to correlate the five supernormal knowledges with objects of five senses. This is an artificial, mechanical interpretation. According to the Vivaraṇa version, *vārtā* refers to "the knowledge of conventional and real truth as they are".[48]

ते समाधावुपसर्गाः व्युत्थाने सिद्धयः ।।37।।

Te samādhāv upasargāḥ vyutthāne siddhayaḥ.

These (supernormal knowledges and powers) are obstacles in meditative trance, but achievements in an extravert state (of mind).

Comments

This is an important aphorism, because it gives a warning to the spiritual practitioner that one should not be tempted towards the supernormal knowledges and powers which one may achieve. But this aphorism is not placed in its proper place, because after this aphorism Patañjali continues to describe different supernormal powers in succeeding aphorisms (YS III. 37–41, 43–45) with a neutral attitude as if no warning was made.

Secondly, it is to be noted in this context that not all supernormal achievements are obstacles on the path towards meditative trance. For, supernormal achievements are of two kinds: material and spiritual. Even the final, or semifinal, state of emancipation is a supernormal achievement. Material achievements can be called obstacles on the path to emancipation, but the spiritual achievements are not such obstacles. The two types of supernormal achievements get mixed up in Patañjali's discussion. Interestingly, such an admixture is found in the Buddhist discussions as well.[49]

Patañjali's criticism of supernormal powers as obstacles in the path of meditative trance is continuation of the Buddha's approach, which is clear in *Dīghanikāya* Sīlakkhandavaggo II.1.1, as Tandon (1998: 79–80) records. According to it, the Buddha warned, "It is therefore not proper to extol such performances. Seeing danger in them, these should rather be treated with contempt, shame and disgust".

Tandon further reports that the Buddha forebade his disciples to exhibit supernormal powers and he declared that if a *bhikkhu* did so, then he would be deemed to have committed an offense of wrongdoing (dukkaṭa).

बन्धकारणशैथिल्यात्प्रचारसंवेदनाच्च चित्तस्य परशरीरावेशः ।।38।।

Bandhakāraṇaśaithilyāt prachārasaṁvedanāc ca cittasya paraśarīrāveśaḥ.

By loosening the causes of bondage and by experiencing the movement (of mind) the mind can enter into the body (of another being).

Comments

I have not found a reference to this type of supernormal power in the Buddhist meditational literature.

उदानजयाज्जलपङ्ककण्टकादिष्वसङ्ग उत्क्रान्तिश्च ।।39।।

Udānajayāj jalapaṅkakaṇṭakādiṣv asaṅga utkrāntiś ca.

By the conquest of the upward vital force (*udāna*), there arises non-sticking with water, mud, thorns etc. and uplift (after death).

Comments

Disappearance of body and passing through water, mud and thorns without sticking to them is accepted in the form of "disappearance" (*antardhāna*) in the Buddhist theory of *ṛddhis*.[50] Patañjali associates it with *udāna*, which is a vital force that is accepted in the Sāṅkhya tradition.[51] *Udāna* is a vital force which works in an upward direction in the body. Patañjali also claims that one attains a higher status after death (*utkrānti*) as one of the effects of mastery over *udāna-vāyu*.

समानजयात्प्रज्वलनम् ।।40।।

Samānajayātprajvalanam.

By establishing mastery over the vital force called *samāna*, arises radiance.

Comments

Samāna is another vital force (*vāyu*) according to Sāṅkhya. It functions in the naval region (*nābhimaṇḍala*) of the body,[52] and the naval region is the location of stomach-fire (*jāṭharāgni*). This is how Patañjali seems to connect *samāna* with fire (radiance).

Asaṅga describes radiance (*jvalana*) as one of the supernormal powers achieved by a *bodhisattva*:

> Bodhisattva who attains the power of radiance, may radiate by the upper part of the body when cool waves of water flow from the lower part. Or

he may radiate by the lower part of the body when cool waves of water flow from the upper part. Or he may meditatively attain the fire element by which he radiates in the whole body. When he radiates by the whole body, the rays of different colors, blue, yellow, red, white, *mañjiṣṭhā*-colour and crystal color are emitted from his body.[53]

श्रोत्राकाशयोः संबन्धसंयमाद्दिव्यं श्रोत्रम् ॥41॥

Śrotrākāśayoḥ saṃbandhasaṃyamād divyaṃ śrotram.

By practicing integrated concentration on the relation between the ether and the organ of hearing, arises divine (supernormal) ear.

Comments

Here Patañjali is acknowledging divine ear (*divyaṁ śrotram*) as one of the supernormal powers. In Buddhist tradition it is accepted as a supernormal cognition (*abhijñā*). Asaṅga has explained the procedure of achieving a divine ear.[54] It does not refer to the connection between ear and ether. Though Patañjali is referring to the same supernormal power, the procedure he recommends is based on the connection between the auditory sense organ and the ether (*ākāśa*). Patañjali's procedure has a metaphysical basis in Sāṅkhya and Vaiśeṣika.

According to Sāṅkhya metaphysics the auditory sense organ and ether are connected through the sound. The sound is cognized through the auditory sense organ, and the essence of sound (*śabda-tanmātra*) is the subtle element which is the cause of ether. In the Vaiśeṣika system, on the other hand, the connection between the ether and the auditory sense organ is more direct. According to it, the auditory sense organ is made of ether. (The ether limited by the physical ear is the auditory sense organ.) They are connected through sound also, because sound, according to Vaiśeṣika, is the quality of ether. Hence, when Patañjali recommends the procedure of concentration on the connection between the auditory sense organ and ether, it is not clear whether he is talking about the connection as accepted by Sāṅkhya or Vaiśeṣika.

Vyāsa, in his commentary, seems clearly to follow Vaiśeṣika when he says:

> Ākāśa is the foundation of all auditory organs and also of all sounds.[55]

That is because, as we have seen, according to the Vaiśeṣika view, the auditory sense organ is the ether covered by ears and sound is the quality of ether.

In both Vaiśeṣika and Sāṅkhya, *ākāśa* (ether) is understood as gross element (*mahābhūta*). It is a positive, material substance. (That is why we translated the term *ākāśa* as "ether" for the sake of convenience.) In Buddhism, however, *ākāśa* is not regarded as a positive substance (like ether) but as "space", and is defined as something without a cover (*anāvaraṇa*).[56] Vyāsa mixes up this Buddhist approach with the Vaiśeṣika and Sāṅkhya approach when he says *ākāśa* is also said to be without cover.[57] Again he associates this description of *ākāśa* (which comes from

Vasubandhu) with Vaiśeṣika descriptions, namely *amūrta* (having unlimited size) and *vibhu* (all-pervasive).[58]

कायाकाशयोः सम्बन्धसंयमाल्लघुतूलसमापत्तेश्चाकाशगमनम् ॥42॥

Kāyākāśayoḥ sambandhasaṅyamāl laghutūlasamāpatteś cākāśagamanam.

As a result of the integrated concentration on the relation between the body and the ether (*ākāśa*), and by meditative attainment of light things like cotton, (one acquires the power of) going through space.

Comments

This aphorism about "going through the sky/space" has at its background the supernormal power of the same kind described by Asaṅga in *Śrāvakabhūmi*.

The procedure of attaining this power, which Patañjali recommends, involves two things: (1) triple concentration on the relation between body and ether and (2) meditative attainment of lightness such as that of cotton.

Asaṅga, while describing this supernormal power, talks about three ideations: (1) lightness of the things such as cotton, (2) softness and (3) space-element. One can attain the power to move in space by intensively resolving oneself as such. With regard to the ideation of lightness, one resolves oneself to be light-weighted such as a piece of cotton (*tūla* or *karpāsa*). The notion of softness supports lightness, and the notion of space-element is that by which one affirms one's own lightness and softness. If such a person wants to go anywhere, then all the interim obstacles to the act of going belonging to a material thing are resolved as "space", by the resolution-based mental application.[59]

Patañjali's statement of the procedure seems to be based on Asaṅga's explanation.

बहिरकल्पिता वृत्तिर्महाविदेहा ततः प्रकाशावरणक्षयः ॥43॥

Bahir arakalpitā vṛttir mahāvidehā tataḥ prakāśāvaraṇakṣayaḥ.

When the state of mind is not constructed as about an external thing, it is called the great disembodied (state). By practicing (integrated concentration) on that, the veil covering light is removed.

Comments

When a mental attitude is without imagination of anything external, then it becomes greatly disembodied. Through that the veil of light is destroyed. This meditative practice is compatible with the Yogācāra view of consciousness, and it also incorporates Jaina soteriology.

According to Yogācāra, particularly its "mind-only" doctrine, the external world is not real but is mentally constructed out of ignorance or misconception. Hence, if the mind stops "constructing" the external would, it stops constructing even the corporeal body and hence becomes greatly disembodied. Patañjali here seems to be using Yogācāra metaphysics only as a practical device.

The idea of *prakāśāvaraṇa*, "the veil covering light", can be appreciated as a reflection of the Jaina concepts of *jñānāvaraṇīya-karma* and *darśanāvaraṇīya-karma* (knowledge-obscuring karma and intuition-obscuring karma). According to Jainism, *jīva* (the analogue of Sāṅkhya's *puruṣa* and Buddhism's *citta*) has capacities like infinite intuition (*ananta-darśana*), infinite knowledge (*ananta-jñāna*). But these capacities are obscured by different types of karma. For instance, the infinite intuition is obscured by the intuition-obscuring karma, and the infinite knowledge is obscured by the knowledge-obscuring karma. Patañjali's notion of *prakāśāvaraṇa* might have been inspired by the Jaina notions of *jñānāvaraṇīya* and *darśanāvaraṇīya-karma*.

There are parallel concepts of *āvaraṇa* (veil or cover) in Buddhism. Vasubandhu talks about a three-fold veil (*āvaraṇatraya*): "Bhagavān states three veils: veil of action (*karmāvaraṇa*), veil of defilements (*kleśāvaraṇa*) and veil of fruition (*vipākāvaraṇa*)" (AKB IV.95cd). Vyāsa, in his interpretation of the term *prakāśāvaraṇa* (the light-obscuring veil) combines the Buddhist notion of three veils with the *triguṇa* theory of Sāṅkhya when he says:

> Due the fixation of mind (called "Greatly disembodied") the three-fold veil of the *sattva* of the cognition, which is essentially the light, gets destroyed. The three-fold veil consists of defilement, action and fruition, and is rooted in *rajas* and *tamas*.[60]

स्थूलस्वरूपसूक्ष्मान्वयार्थवत्त्वसंयमाद्भूतजयः ।।44।।

Sthūlasvarūpasūkṣmānvayārthavattvasaṁyamād bhūtajayaḥ.

By integrated concentration on the gross form, essential nature, subtle form, relation and purposefulness, one obtains mastery over elements (*bhūta*).

Comments

According to Vyāsa's interpretation, every *bhūta* (element: earth, water, fire, air and ether) has five aspects:

(1) *Sthūla*: gross.
(2) *Svarūpa*: its own character. According to Vyāsa, it is two-fold: *sāmānya* and *viśeṣa*. Here Vyāsa defines *dravya* as *sāmānya-viśeṣa-samudāya*. Then, he talks about two types of "collections", non-composite (discretely formed group) and composite (which can be called *avayavī*). Here Vyāsa introduces another conception of *dravya*: the collection which is inseparably related to its parts (*ayutasiddha-avayava*) is *dravya*.
(3) *Sūkṣma*: subtle. According to Sāṅkhya, *tanmātra* is the subtle form of the *bhūtas*. Vyāsa refers to that and also to the smallest particles (*paramāṇus*) from which *bhūtas* are constituted. Hence, he mixes up Sāṅkhya and Vaiśeṣika ideas of subtle *bhūtas*.
(4) *Anvaya*: By *anvaya*, Vyāsa refers to three *guṇas* (strands of *prakṛti*) which run through all *bhūtas*.

(5) *Arthavattva*: purposiveness. Here again Vyāsa refers to the Sāṅkhya teleology according to which *prakṛti* and its manifestations are active for the sake of *puruṣa*.

In the Buddhist theory of meditation, concentration on whole-objects (Pali: *kasiṇas*, Sanskrit: *kṛtsna*), often termed as *kṛtsnāyatana*, consists of the meditation on ten objects: four gross elements, *ākāśa*, *vijñāna* and four colours. Concentration on *kasiṇas* develops in stages from gross to subtle forms and ultimately results into control over the gross elements. "*Kṛtsnāyatanas* cause entrance into *abhibhvāyatana* (mastery)".[61]

The aphorism by Patañjali can be understood as a reformulation of the idea of meditation on *kasiṇas*.

Vyāsa's definitions of *bhūtas* (particularly those of water, fire and ether) are similar to Vasubandhu's definitions:

(1) Earth

Vasubandhu: *Kharaḥ pṛthvīdhātuḥ* + *pṛthivī varṇa-saṁsthānam* (=colour+structure)
Vyāsa: *Mūrtir bhūmiḥ*

(2) Water

Vasubandhu: ***Snehaḥ*** *abdhātuḥ*
Vyāsa: ***Sneho*** *jalam*

(3) Fire

Vasubandhu: ***Uṣṇatā*** *tejodhātuḥ*
Vyāsa: *Vahnir* ***uṣṇatā***

(4) Air

Vasubandhu: *Īraṇā vāyudhātuḥ*
Vyāsa: *Vāyuḥ praṇāmī*[62]

(5) Ether/space

Vasubandhu: *Āvaraṇasvabhāvam ākāśaṃ* ***yatra rūpasya gatiḥ***
Vyāsa: ***Sarvatra gatir*** *ākāśaḥ*

Vyāsa presents two notions of *dravya*: (1) *sāmānyaviśeṣasamudāyo dravyam* (*Dravya* is a collection of general and specific characteristics.) (2) *ayutasiddhāvayavaḥ samūho dravyam iti patañjaliḥ* (*Dravya* is a collection (whole) which is inseparably related to its parts.)

The first definition is closer to Buddhism, the second is influenced by the Vaiśeṣika concept of *avayavin*.

There is close similarity between Patañjali's methods of concurring *bhūtas* and those of concurring *indriyas*. For discussion, see YS III.47.

ततोऽणिमादिप्रादुर्भावः कायसम्पत्तद्धर्मानभिघातश्च ।।45।।

Tato'ṇimādiprādurbhāvaḥ kāyasampat taddharmānabhighātaś ca.

Thence there is arising of atomic size (aṇimā) and so on as also the bodily wealth and non-resistance by their characteristics.

Comments

In the orthodox tradition, the supernormal powers are often presented as eightfold. In the early Buddhism literature, we do not find them in this form, but all of the eight types are found separately in the Buddhist theory of *abhijñā*s and *ṛddhis* (supernormal knowledges and powers). Vyāsa's exposition of the eight powers closely resembles Asaṅga's explanations in *Śrāvakabhūmi*[63] as shown in Table 3.9.

रूपलावण्यबलवज्रसंहननत्वानि कायसंपत् ।।46।।

Rūpalāvaṇyabalavajrasaṃhananatvāni kāyasaṃpat.

The bodily wealth consists in beauty, grace, strength and thunderbolt-like hardness.

Comments

In the preceding aphorism, Patañjali stated three results of the conquest of material elements: (1) the supernormal powers such as *aṇimā*, (2) the bodily wealth and

TABLE 3.9 Eight supernormal powers

S.N.	*Name of the power*	*Vyāsa's description*	*Asaṅga's explanation*
1	*Aṇimā*	One becomes minute.	By the notion of resolution a gross thing can be minute.
2	*Laghimā*	One becomes light.	By concentrating on lightness one becomes light.
3	*Mahimā*	One becomes large.	A minute thing becomes big.
4	*Prāptiḥ*	One touches the moon by his fingertip.	One touches sun and moon by hand.
5	*Prākāmyam*	(Non-frustration of desire) One dives in and comes out of earth like in water.	One dives in the ground and comes out of it as if through water.
6	*Vaśitvam*	One controls material things and does not let them control him/her.	Controls the worlds up to Brahmaloka.
7	*Īśitṛtvam*	One is capable of creating and destroying compound things.	One transforms oneself into many and again many into oneself.
8	*Yatrakāmāvasāyitvam*	Fulfilment of desire (*satyasaṅkalpatā*).	One changes things according to one's resolve (*adhimokṣa*).

Sources: YB III.45; ŚB: 460–68

(3) non-resistance by the properties of the material elements. In this aphorism, he is explaining what he includes under the bodily wealth.

His explanation seems to echo the description of Buddha's physiological wealth as given by Vasubandhu. Vasubandhu says that all the Buddhas which arise in different periods have the following three kinds of wealth in common:[64]

(1) *Hetu-sampat* (causal wealth)
(2) *Phala-sampat* (resultant wealth)
(3) *Upakāra-sampat* (favour done to the world as wealth)

Under resultant wealth there are four kinds:

(2a) *Jñāna-sampat*: the wealth of knowledge
(2b) *Prahāṇa-sampat*: the wealth of abandonment
(2c) *Prabhāva-sampat*: the wealth of influence
(2d) *Rūpakāyasampat*: the wealth of the physical appearance and body

Here we are concerned with (2d) *Rūpakāyasampat*. The Buddha has beautiful and strong bodily form. This is supposed to be the result of accumulation of merit. Vasubandhu describes *Rūpakāyasampat* as follows: "2d (1) *Lakṣaṇa-sampat*, 2d (2) *Anuvyañjana-sampat*, 2d (3) *Bala-sampat* and (4) *Vajrasārāsthi-sampat*. It is easy to see how Patañjali's description of *kāyasampat* reflects Vasubandhu's description of the Buddha's bodily beauty and strength as shown in Table 3.10.

This is one of the places where Patañjali has tried to appropriate the tallest claims of Buddhism as achievable by practicing yoga as conceived by him.

TABLE 3.10 Bodily wealth and beauty as a supernormal achievement

S.N.	*Vasubandhu's description of* Rūpakāyasampat	*Patañjali's description of* Kāyasampat
1	*Lakṣaṇasampat* This consists of thirty-two major signs of bodily grandeur and beauty.	Patañjali simply talks about *rūpa-lāvaṇya*, that is beauty of the appearance without mentioning specific signs.
2	*Anuvyañjanasampat* This consists of eighty secondary signs of bodily beauty.	
3	*Bala-sampat* The wealth of bodily strength.	*Bala* (strength)
4	*Vajrasārāsthisampat* The wealth of thunderbolt-like (unbreakable) bones.	*Vajrasaṁhananatva* Thunderbolt-like firmness of body.

Sources: AKB VII.34, YS III.46

ग्रहणस्वरूपास्मितान्वयार्थवत्त्वसंयमादिन्द्रियजयः ।।47।।

Grahaṇasvarūpāsmitānvayārthavattvasaṃyamādindriyajayaḥ.

As a result of integrated concentration on grasping, its nature, I-notion, relation and purposiveness, arises the conquest of the sense-organs.

Comments

This aphorism follows the same pattern as III.43.

The parallelism can be seen by Table 3.11.

Two aphorisms represent two routes of creation, according to the Sāṅkhya process of creation.

Bhūta route

Buddhi—ahaṁkāra—sūkṣma (tanmātra)—bhūta

Indriya route

Buddhi—ahaṁkāra (asmitā)—indriya - and-*manas*

One of the strategies Patañjali applies is that in order to concur something we have to focus on what it arises from and how it arises.

Supposing that *bhūta*s are to be concurred, we notice that *bhūta*s as gross entities arise from "subtle *bhūta*s", that is *tanmātra*s. And they arise from them through their internal relation and through their goal-orientedness (that is, as they are active for the sake of *puruṣa*). So, one can concur *bhūta*s by focusing on the nature of the *bhūta*s, their causes, namely *tanmātra*s, their internal connectedness in terms of three *guṇa*s and their purposiveness.

Similarly, now supposing that *indriya*s are to be concurred: we notice that *indriyas* grasp their objects (*grahaṇa*) because they arise from *ahaṁkāra* (same as *asmitā*) through their internal connection in terms of the three *guṇa*s and also their orientedness towards the goals of the *puruṣa*.

ततो मनोजवित्वं विकरणभावः प्रधानजयश्च ।।48।।

Tato manojavitvaṁ vikaraṇabhāvaḥ pradhānajayaś ca.

Thence arise mind-like speediness, capacity to transform things and mastery over the first cause (*pradhāna*).

TABLE 3.11 The common pattern followed in two aphorisms

III.44	*Sthūla*	*Svarūpa*	*sūkṣma*	*anvaya*	*arthavattva*	*saṁya-māt*	*Būta-jayaḥ*
III.47	*grahaṇa*	*Svarūpa*	*asmitā*	*anvaya*	*arthavattva*	*Saṁya-mād*	*Indriya-jayaḥ*

Sources: YS III.44 and 47

Comments

Here Patañjali is stating three advantages of the conquest of sense-organs:

(1) ***Manojavitva*** (capacity to move with mind-like speed): According to Buddhism, only the Buddha has this type of speed. Patañjali makes it applicable to all yogins. Perhaps being aware of the exaggeration in this claim, Vyāsa interprets *manojavitvam* as "*anuttamo gatilābhaḥ*" (that is, attainment of excellent speed). In Vasubandhu's framework, an excellent speed can be attained by yogins in general.
According to Vasubandhu's classification, supernormal power is of two kinds: the power for motion (*gati*) and the power for creation (*nirmāṇa*). The power for motion is the capacity to move fast. Vasubandhu divides this speedy movement into three kinds: (1) *vāhinī gati*, the capacity to move one's body like a bird; (2) *ādhimokṣikī gati*, the capacity to move body anywhere by making a resolve; and (3) *manojavā gati*, mind-like speed. Vasubandhu says that only the Buddha has this type of speed.[65]

(2) ***Vikaraṇabhāva***: This term is quite problematic. Its literal meaning is "being without *karaṇa* (that is, without *indriya*)". Vyāsa also has found it difficult to interpret. He splits the word into *Vi* (=*videha*) + *karaṇa* (*indriya*) and interprets the word artificially. I suggest that the original word could be different, such as "*vikurvaṇabhāvaḥ*". This term has a relevant meaning. According to Edgerton's Buddhist Hybrid Sanskrit Dictionary, *vikurvaṇa* = miracle-maker. Hence *vikurvaṇa-bhāva* would mean "being in a position to play miracles". This meaning would perfectly fit into the context. And even if we accept the present reading, it can be interpreted as being in a position to transform (*vi*+*karaṇa* = to make otherwise, to transform) things.

(3) ***Pradhānajaya***: Mastery over primordial nature. This is a supernormal power explained in Sāṅkhya terms. Here, as in many other places, Patañjali seems to present Buddhist and Sāṅkhya ideas side by side.

सत्त्वपुरुषान्यताख्यातिमात्रस्य सर्वभावाधिष्ठातृत्वं सर्वज्ञातृत्वं च ।।49।।

Sattvapuruṣānyatākhyātimātrasya sarvabhāvādhiṣṭhātṛtvaṃ sarvajñātṛtvaṃ ca.

The one who recognizes the distinction between mind and spirit attains supremacy over all states of being and omniscience.

Comments

As suggested in the commentary on YS III.37, Patañjali and also Buddhist philosophers discuss the two kinds of supernormal achievements, namely material and spiritual, as on par with each other. For example, in the list of supernormal knowledges (*abhijñā*), in Buddhism *āsravakṣayajñāna* (the knowledge of the form "all intoxicants are destroyed"), which is the liberating knowledge, is included along with the supernormal knowledges, which are more of material nature. Similarly, in the aphorism Patañjali includes attainment of Sāṅkhya-type liberating knowledge in the list of supernormal powers (*siddhi, vibhūti*). According to Sāṅkhya tradition, the discrimination between *prakṛti* and *puruṣa* liberates. In Patañjali's reformulation of practical Sāṅkhya, it takes the form of the discrimination between mind and spirit (*sattvapuruṣānyatākhyāti*). Because of the metaphysical character of this knowledge, its results are also metaphysical. Vyāsa explains these results "*sarva-bhāvādhiṣṭhātṛtvam*" (becoming the abode of all that is there) and *sarvajñātṛtvam* (knowledge of everything) in terms of Sāṅkhya.

In Buddhism where six supernormal powers (*abhijñās*) are stated, the first five powers are mundane (*laukika*), whereas the sixth one (namely knowledge of the form, "Intoxicants are destroyed": *āsravakṣayajñāna*) is supra-mundane, which only an *arhat* (liberated person) has. Similarly, Patañjali is here including supra-mundane powers such as omniscience, discriminatory knowledge and so on. In the next aphorism he will talk about final liberation as the supernormal achievement.

तद्वैराग्यादपि दोषबीजक्षये कैवल्यम् ।।50।।

Tadvairāgyād api doṣabījakṣaye kaivalyam.

By being detached even to that, one attains isolated-ness (Kaivalya) when the very seed of defectiveness is destroyed.

Comments

Patañjali has accepted the soteriological framework spelled out by Buddhist thinkers such as Asaṅga and Vasubandhu, according to which, in order to attain a higher plain in one's spiritual journey, one has to develop detachment towards the lower planes (*bhūmis*). The Buddhists explain these planes in terms of the three *dhātus*, namely *kāmadhātu, rūpadhātu* and *ārūpyadhātu*. When one reaches *ārūpyadhātu*, one has to develop detachment to all objects belonging to *kāma* and *rūpa*. But one could be attached to *ārūpyadhātu* in this stage. In order to attain final liberation (namely *saṁjñā-vedayita-nirodha*), one has to develop detachment towards formless objects (that is, objects belonging to *ārūpyadhātu*) as well. Asaṅga calls this "*lokottareṇa mārgeṇa vairāgyam*", which is the detachment towards objects belonging to all three *dhātus*.

Patañjali replaces the Buddhist hierarchy of worlds and objects belonging to them by the Sāṅkhya hierarchy of *guṇa*s and their manifestations. Accordingly, he distinguishes between lower and higher *vairāgya* (in YS I.15–16). The same distinction is being applied here.

The metaphors of seed and sprout for explaining the basic defilements as the cause of bondage/transmigration and those of destruction of seed and destruction of sprout for explaining cessation of defilements and attainment of liberation are very much central to Buddhism. As Vasubandhu says:

Therefore, the transmigration (*saṁsāra*) has no beginning. But the end does occur due to the destruction of the cause; because the birth depends on cause, like the end of sprout, which occurs due to the destruction of the seed.[66]

स्थान्युपनिमंत्रणे सङ्गस्मयाकरणं पुनरनिष्टप्रसङ्गात् ।।51।।

Sthānyupanimantrane saṅgasmayākaraṇaṃ punar aniṣṭaprasaṅgāt.

When (the yogin in a high stage) is invited by deities holding different positions, he does not become attached or proud as he thinks that undesirable consequences will follow again.

The idea in this aphorism is that gods invite the yogin who has attained a high spiritual stage; but a true yogin does not develop any attachment or pride towards that, thinking that developing such an attitude will again have undesirable consequences. The specific source of this idea is not known. The general idea is well taken. And it is that at every high position there are new temptations and new reasons for becoming proud—which are in fact the dangers that should be avoided.

In the commentary, Vyāsa refers to four types of yogins. This seems to be an expansion of the three-fold classification of *yogācāras* (yoga-practitioners) given by Asaṅga[67] and Vasubandhu.[68] This is clear from Table 3.12.

Vyāsa here seems to have slightly expanded and reframed the Asaṅga–Vasubandhu classification.

क्षणतत्क्रमयोः संयमाद्विवेकजं ज्ञानम् ।।52।।

Kṣaṇatatkramayoḥ saṃyamād vivekajaṃ jñānam.

By practicing integrated concentration on moments and their sequence, arises discriminative knowledge.

TABLE 3.12 Classification of yoga-practitioners according to the two traditions

Asaṅga and Vasubandhu classification of yoga-practitioner	*Vyāsa's classification of yogins*
First stage: *Ādikarmika* (beginner) Second stage: *Kṛtaparijaya* (who has mastered the method) or *kṛtaparicaya* (experienced practitioner) Third stage: *Atikrāntamanaskāra* who has transcended the stage of mental application	First stage: *Prathama-kalpika* who has started and got a glimpse Second stage: *Madhubhūmika* who has attained *ṛtambharā prajñā* Third stage: *Bhūtendriya-jayī* who has mastered *bhūtas* and *indriyas* Fourth stage: *Atikrāntabhāvanīya* who has transcended the stage of the need for meditative practice, who has attained seven-fold ultimate wisdom (*prāntabhūmi prajñā*)

Sources: ŚB 284–285, AKB VI.10–11, YB III.51

Comments

This is a perplexing aphorism, because the question is, how moments and the sequential order among them are relevant to *vivekaja-jñāna* (the knowledge arising from discrimination). That is because *vivekaja-jñāna* pertains to the distinction between *prakṛti* and *puruṣa* or *sattva* and *puruṣa* (that is, *buddhi* and *puruṣa*) as Patañjali maintains elsewhere. But none of them are momentary.

Developing concentration on moments and their sequential order appears very much Buddhistic. But even in the mainstream Buddhist tradition this type of meditational practice is not found.

When I searched in AKB, I found that Vasubandhu is talking of the ordered sequence of moments giving rise to liberation, when he talks about the sequence of sixteen consciousness-moments (*citta-kṣaṇa*s) which constitute an ordered sequence of direct realizations (*abhisamaya-krama*). The sixteen *citta*s are stated in Table 3.13.

It is not necessary that Patañjali would have accepted all this in detail. But he seems to be following the core idea that while developing spiritually, one experiences consciousness-moments of spiritual development in a certain order. Concentrating on this developmental order may lead to final liberation. When Patañjali is talking about moments and sequential order among them, he is talking of this type of consciousness moments (*citta-kṣaṇa*) and not moments in general.

How to interpret *vivekajaṁ jñānam* (knowledge arising from *viveka*) is another question in this context. The Sāṅkhya concept of *viveka* (discrimination) between

TABLE 3.13 Ordered sequence of direct realizations (*abhisamaya-krama*)

Consciousness moment no.	*Realm*	*Āryan truth*	*Patient acceptance* (kṣānti) *or knowledge*
1	Sensuous	Suffering	Patient acceptance of the law
2	Sensuous	Suffering	Knowing the law
3	Form and formlessness	Suffering	Subsequent patient acceptance
4	Form and formlessness	Suffering	Subsequent knowledge
5	Sensuous	Origination	Patient acceptance of the law
6	Sensuous	Origination	Knowledge of the law
7	Form and formlessness	Origination	Subsequent patient acceptance
8	Form and formlessness	Origination	Subsequent knowledge
9	Sensuous	Cessation	Patient acceptance of the law
10	Sensuous	Cessation	Knowledge of the law
11	Form and formlessness	Cessation	Subsequence patience acceptance
12	Form and formlessness	Cessation	Subsequent knowledge
13	Sensuous	Path	Patient acceptance of the law
14	Sensuous	Path	Knowing the law
15	Form and formlessness	Path	Subsequent patient acceptance
16	Form and formlessness	Path	Subsequent knowledge

Source: AKB VI.25–26

two simultaneously existing entities such as *prakṛti* and *puruṣa* is not applicable here.

Here there is a question of *viveka* in relation to two consciousness moments in sequential order. Vasubandhu uses the expression *vivekaja* in the sense of "that which arises from dissociation" when he talks of the stages of spiritual development from lower *dhyāna* to higher *dhyāna*. As Vasubandhu says:

> By dissociating from the fourth *(rūpa-)dhyāna*, arises the sphere of infinite space. By dissociating from it arises the sphere of infinite consciousness. By dissociating from it arises the sphere of nothingness. By dissociating from it arises the sphere of neither perception nor non-perception. In this way (arise) four formless spheres. What is this dissociation (*viveka*)? It is the way by which one is freed from the lower plane. (This is done by) developing detachment.[69]

This throws light on the possible meaning of *vivekajaṁ jñānam* in this aphorism. It means the knowledge arising by dissociation from the lower plane through detachment.

I think this is one of the locations where Patañjali freely thinks in a Buddhist way.

However, Vyāsa has a problem here because if one talks about sequential order among moments without continuity, then it amounts to leaving the Sāṅkhya framework. In order to avoid this problem, he insists that "Whatever are the past and future moments, they are to be explained as connected through transformation (*pariṇāma*) (of a thing)".[70]

जातिलक्षणदेशैरन्यतानवच्छेदात्तुल्ययोस्ततः प्रतिपत्तिः ।।53।।

Jātilakṣaṇadeśair anyatānavacchhedāt tulyayos tataḥ pratipattiḥ.

From this arises the distinct knowledge of (any) two similar things which cannot be distinguished by class, characteristic marks and position in space.

Comments

Here Patañjali is stating the advantage of the *vivekaja-jñāna*, which he stated in the earlier aphorism. The yoga-practitioner who concentrates on consciousness-moments occurring in sequential order is able to recognize two moments which are otherwise similar, but temporally different, when because of the similarity the difference between them cannot be made out in terms of type, characteristics and spatial location.

तारकं सर्वविषयं सर्वथाविषयमक्रमं चेति विवेकजं ज्ञानम् ।।54।।

Tāraka sarvaviṣayaṁ sarvathāviṣayam akramaṁ ceti vivekaja jñānam.

The discriminative knowledge is the liberating one, has all things as its objects. It knows objects with all their (temporal) characteristics; it is (one) without sequence (of elements).

Comments

Vivekajaṁ jñānam, as we have seen, can be understood in two ways. (1) The cognition which discriminates between *buddhi* and *puruṣa* (Sāṅkhya version) and (2) the cognition of higher plane which arises by dissociating from (*viveka*) the lower plane through detachment.

In both the senses, the discriminating (or separating) knowledge is supposed to be liberating (*tāraka*), it encompasses all objects (*sarva-viṣaya*), it observes objects in all its aspects (*sarvathāviṣayam*) and (although it takes into account sequence among momentary consciousnesses) in itself it is not sequential.

It is not clear why Vyāsa interprets the word *tāraka* as that which is attained through one's own intuition, and which is not based on instruction (*svapratibhottham anaupadeśikam*), when the word simply means "savior"—that which takes to the other shore; that which liberates.[71]

सत्त्वपुरुषयोः शुद्धिसाम्ये कैवल्यम् ।।55।।

Sattvapuruṣayoḥ śuddhisāmye kaivalyam.

When the mind (*sattva*) and the spirit (*puruṣa*) become equal in purity, there is isolation.

Comments

This is a purely a Sāṅkhya explanation of liberation (*kaivalya*). According to Sāṅkhya, one is liberated when *buddhi* which is dominated by *sattvaguṇa* becomes so purified that it becomes very similar to *puruṣa*. So "when there is similarity of purity between *buddhi* and *puruṣa*" means "when *buddhi* becomes as pure as *puruṣa*". Vyāsa, too, explains the situation by using Sāṅkhya terminology.

Notes

1 "*nābhicakre, hṛdayapuṇḍarīke, mūrdhni jyotiṣi, nāsikāgre, jihvāgre ityevamādiṣu deśeṣu bāhye vā viṣaye cittasya vṛttimātreṇa bandha iti dhāraṇā*" (YB III.1) [*Dhāraṇā* is fixing the mind, by its mere presence in the objects such as the naval circle, the heart-lotus, the light in the top of the head, the tip of the nose and the tip of the tongue or in an external object.]

2 These features of *samādhi* are shared by *nirvitarkā samāpatti*. See the comments on YS I.43.

3 "*tṛtīye pañca tūpekṣā smṛtiḥ prajñā sukhaṁ sthitiḥ*" (AKB VIII.8).

4 "*ānāpānasmṛtiḥ prajñā pañcabhūḥ*" (AKB VI.12).

5 "*Svādhobhūviṣayāḥ*" (AK VII.44). Also see AKB.

6 "*cittacaittā acaramā utpannāḥ samanantaraḥ*" [Minds and mental states which arise, other than the ultimate mind (of *arhat*) are the immediate cause.] (AK II.62ab).

7 "*samanantarapratyayas tad yathā nirodhasamāpatticittaṁ vyutthānacittasya*" (AKB II.62ab, p. 98).

8 "*nirodhasamāpattivyutthitaḥ paraṁ cittaśāntiṁ labhate. nirvāṇasadṛśatvāt samāpatteḥ*" (AKB IV.56).

9 Vasubandhu says this clearly, "*keyam ekāgratā nāma? ekālambanatā. evaṁ tarhi cittāny evaikālambanāni samādhir na cetasikaṁ dharmāntaram iti prāpnoti. Na cittāny eva samādhiḥ. Yena tu tāny ekāgrāṇi vartante sa dharmaḥ samādhiḥ saiva cittaikāgratā*" [What is this single-pointedness? It means having a single object. (Objection:) But in that case minds which

have a common object will be identical with *samādhi* (meditative absorption). Hence *samādhi* will not be the separate character namely a mental factor. (Answer:) Minds themselves are not identical with *samādhi*. *Samādhi* is the characteristic due to which they become single-pointed. The same is the single-pointedness of mind.] (AKB VIII.1d). This clearly suggests that commonness of the object makes the concentrated cognitions similar.

10 Here I am translating the term *lakṣaṇapariṇāma* as temporal characteristic or temporal dimension, following Bhadanta-Ghoṣaka's view as reported by Vasubandhu in AKB V.25. Vyāsa in his commentary has accepted this interpretation when he says, "*lakṣaṇapariṇāmaś ca nirodhas trilakṣaṇas tribhr adhvabhir yuktaḥ*" [The transformation in terms of *lakṣaṇa* means that the cessation has three *lakṣaṇa*s; that is, it is qualified by three temporal dimensions] (YB III.13).

11 The connection of this aphorism, the next aphorism (YS III.14) and a few others (for example, YS IV.12) with Sarvāstivāda of Buddhism has been noticed by many scholars, such as Stcherbatsky (1961: 37–38), Frauwallner (1984: 321–328) and La Vallée-Poussin (1937).

12 "*yathā caikatve'pi strī mātā cocyate duhitā ca svasā ceti*" (YB III.13) compare "*yathaikā strī mātā vocyate duhitā veti*" (AKB V.26ab, p. 297).

13 "*dvitīyasyādhvasaṅkaraḥ prāpnoti*" (AKB V.26ab, p. 297). Compare this with "*atra lakṣaṇapariṇāme sarvasya sarvalakṣaṇayogād adhvasaṅkaraḥ prāpnotīti parair doṣaś codyate iti tasya parihāraḥ*" (YB III.13).

14 "*kiṁ kāritrasyāpy anyad asti kāritram / atha tannaivātītaṁ nāpyanāgataṁ na pratyutpannamasti ca / tenāsaṁskṛtatvān nityamastīti prāptam*" (AKB V.27, p. 298). Vyāsa refers to this objection as "*lakṣaṇānām avasthānāṁ ca kauṭasthyaṁ prāpnoti*" (YS III.13). (*kauṭasthya* is the same as *nityatva*).

15 "*guṇi-nityatve'pi guṇānāṁ vimardavaicitryāt*" (YB III.13).

16 Vasubandhu while criticizing *pariṇāmavāda* also criticizes the Sāṅkhya concept of *dharmī*: "*kathaṁ ca sāṅkhyānāṁ pariṇāmaḥ? avasthitasya dravyasya dharmāntaranivṛttau dharmāntaraprādurbhāva iti. kaś cātra doṣaḥ? sa eva hi dharmī na saṁvidyate yasyāvasthitasya dharmāṇāṁ pariṇāmaḥ kalpyeta*" [What is the nature of transformation (*pariṇāma*) according to Sāṅkhyas? It is the rise of a different property in an enduring substance when it's one property ceases. What is the fault in this view? The property-bearer itself is not (distinctly) cognized, during the endurance of which its properties are supposed to be transformed.] (AKB III.50a).

17 "*yasya tu dharmamātram evedaṁ niranvayaṁ tasya bhogābhāvaḥ*" (YB III.14).

18 According to Abhidharma, *dharma* is that which has its own nature (*svalakṣaṇadhāraṇād harmaḥ*).

19 "*kathaṁ tarhi bhagavān jānāti amuṣyānāgatasyānantaram idam anāgataṁ bhāvīti*" (AKB II.62ab, p. 99).

20 "*atītasāmpratānumānāt*" (AKB II.62ab, p. 99).

21 The term *dharma* here can mean "thing/phenomenon" or "good conduct/meritorious action".

22 "*evaṁ tarhi bhagavān pūrvāntam adṛṣṭvā' parāntaṁ na jānīyāt*" (AKB II.62 (p. 99)).

23 "*tasmāt sarvam icchāmātreṇa bhagavān jānātīti sautrāntikāḥ*" (AKB II.62 (p. 99)).

24 It is interesting to note that more or less in the same period, Bhartṛhari, in his *Vākyapadīya*, builds his theory of language based on such an admixture. He claims: "It is the essence of word, which is Brahman without beginning and end, from which arises the world-process, as manifestation of its meaning" (VP I.1) and also "There is no cognition without association with a word" (VP I.123).

25 (ŚB: 464–465). See Appendix II for details.

26 (ŚB: 466–467). See Appendix II for details.

27 "*kiṁ paracittajñānaṁ paracittasyākāram ālambanaṁ vā gṛhṇāti? na gṛhaṇāti, ākārālambananirapekṣaṁ tad raktam idaṁ cittam iti jānāti na tu amuṣmin rūpe raktam iti jānāti. Anyathā hi tad rūpālambanam api syāt. tadālambanaṁ ca paracittaṁ gṛhaṇataḥ svabhāvagrahaṇaṁ prāpnuyāt*" (AKB VII.11d, (p. 398)). La Vallée-Poussin (1937) has noticed this.

28 See Appendix II for details.
29 Tandon (1998: 23) refers to Buddha's statement from *Dīghanikāya* (*Mahāvagga*), which indicates that the Buddha forecast his death three months before.
30 "*viśuddhāni catvāri dhyānāni niśritya anyatīrthīyaḥ śrāvako bodhisattvo vā abhinirharati catvāri apramāṇāni evaṁ pañcābhijñāḥ*" (ADS: 99).
31 "*pāpaśīleṣūpekṣā na tu bhāvanā. tataś ca tasyāṁ nāsti samādhir ityato na balam upekṣātas tatra saṁyamābhāvāt*" (YB III.23).
32 According to Edgerton (Buddhist Hybrid Sanskrit Dictionary), *gandhahastin* is "an odorous elephant", an elephant in the climax of must.
33 According to Edgerton (Buddhist Hybrid Sanskrit Dictionary), *mahānagna* is a corrupt form of the Pali word *mahānāga* ("a great elephant").
34 According to Edgerton (Buddhist Hybrid Sanskrit Dictionary), *praskandin* means violent or insolent. In this context this seems to be the qualification of an elephant.
35 According to Edgerton (Buddhist Hybrid Sanskrit Dictionary), *varāṅga* is an adjective used for elephants and other parts of the army.
36 According to the *Mahābhārata* story, Cāṇūra, a powerful wrestler, was in Kaṁsa's army. It is said he was defeated and killed by Krishna.
37 "*nārāyaṇabalaṁ kāye sandhiṣv anye daśādhikam / hastyādi saptakabalam*" [The Buddha has Nārāyaṇa power in his body. According to some he had that power in each joint. It is the power of seven (great beings) starting from an elephant and adding ten each time.] (AKB VII.31). Vasubandhu explains, "The power of ten ordinary elephants is equal to the power of one Odorous elephant (*Gandhahastin*)". In this way adding "ten times" for the each succeeding type, you get the powers of the great elephant (*Mahānagna*), insolent elephant (*Praskandin*), a great-limbed elephant (*Varāṅga*), (the giant called) Cāṇūra and Nārāyaṇa (someone parallel to lord Viṣṇu?).
38 "*atidūrāt sāmīpyād indriyaghātān mano'navasthānāt/saukṣmyād vyavadhānād abhibhavāt samānābhihārāc ca //*" (SK 7).
39 I am not saying that this is a scientific model or that Patañjali is applying it scientifically. I am also not saying that this model is influenced by Buddhism. The model, I think, throws light on Patañjali's mode of occult thinking.
40 "*śiṣyasādhāraṇā anye dharmāḥ kecit pṛthagjanaiḥ/araṇāpraṇidhijñānapratisaṁvidguṇādayaḥ //*" (AK VII.35).
41 "*idaṁ jānīyām*" (AKB VII.37).
42 *navajñānasvabhāvā pratibhāna-saṁvid anyatra nirodhajñānāt* (AKB VII.39).
43 "*sarvabhūḥ*" (AKB VII.39).
44 "*āsāṁ kila pratisaṁvidāṁ gaṇitaṁ, buddhavacanaṁ, śabdavidyā hetuvidyā ca pūrvaprayogo yathākramam*" (AKB VII.40).
45 See the entry on the term *hadaya*, Davids and Stede: Pali English Dictionary.
46 "Through *prātibhā-jñāna* (intuitive knowledge), everything can be known" (YS III.32).
47 Vyāsa calls it divine taste (*divyarasa*) (YB III.36).
48 "*vārtātaḥ saṁvyavahāratattvarūpaṁ yathāvad adhigacchati*" (YS(V) III.36).
49 For Patañjali's mention of spiritual achievements as supernormal achievements, see YS III.5, 35, 46, 48, 49, 51, 53, 54. In Buddhist tradition, many supernormal cognitions and powers are material gains. But *āsavakṣayajñāna* (knowledge of destruction of intoxicants) is a spiritual achievement.
50 See Appendix II.
51 According to Sāṅkhya, the mind (that is, the internal organ, *antaḥkaraṇa*) consists of three parts: cognition (*buddhi*), ego (*ahaṁkāra*) and conception (*manas*). Each one has its distinctive role to play. They also perform a collective or common role. This common function of tripartite mind is the five vital forces or wind powers. ("*sāmānyakaraṇavṛttiḥ prāṇādyā vāyavaḥ pañca*" (SK 29)).
52 According to Sāṅkhya and Vedic tradition there are five major vital forces working in different parts of the body. The names of the vital forces and their fields of operation are *prāṇa* (heart), *apāna* (anus), *samāna* (naval circle), *udāna* (throat) and *vyāna* (the whole body).

53 "*tatra jvalanam. ūrdhvaṁ kāyāt prajvalati. adhaḥ kāyāt śītalā vāridhārāḥ syandante. adhaḥ kāyāt prajvalati. uparimāt kāyāt śītalā vāridhārāḥ syandante. tejodhātumapi samāpadyate. sarvakāyena prajvalati. sarvakāyena prajvalitasya vividhā arciṣaḥ kāyān nirgacchanti nīlapītalohitāvadātamañjiṣṭhāḥ sphaṭikavarṇāḥ*" (BSB: 41). According to Apte's Practical Sanskrit English Dictionary, *mañjiṣṭhā* colour is the colour of Indian madder.

54 See Appendix II.

55 "*sarvaśrotrāṇām ākāśaṁ pratiṣṭhā sarvaśabdānāṁ ca*" (YB III.41).

56 "*tatrākāśam anāvṛtiḥ*" "*anāvaraṅasvabhāvam ākāśaṁ yatra rūpasya gatiḥ*" (AK and AKB I.5).

57 "*anāvaraṇaṃ coktam*" (YB III.41).

58 "*tathā mūrtasyānyatrākāśādāvaraṇadarśanād vibhutvam api prakhyātam ākāśasya*" [Since things with limited size (*mūrta*) other than ether are seen to be covered, the all-pervasiveness (*vibhutva*) of ether is well-known.] (YS(V) III.41).

59 See Appendix II.

60 "*tataś ca dhāraṇātaḥ prakāśātmano buddhisattvasya yad āvaraṇaṁ kleśakarmavipākatrayaṁ rajastamomūlaṁ tasya kṣayo bhavati*" (YB III.43).

61 "*abhibhvāyatanapräveśikāni kṛtsnāyatanāni*" (AKB VIII.36d).

62 The meaning of the word *praṇāmī* is not clear. Literally, it means that which bows down. Frauwallner (1984: 281) translates *praṇāmitā* as forward movement.

63 See Appendix II.

64 See AK and AKB VII.34. La Vallée-Poussin (1937) has noticed this.

65 "*śāstur manojavā (gatiḥ)*" (AKB VII.48c). La Vallée-Poussin (1937) has noticed this.

66 "*tasmān nasty eva saṁsārasyādiḥ. antastu hetukṣayād yuktaḥ. Hetvadhīnatvāj janmano bījakṣayād ivāṅkurasyeti*" (AKB III.19d).

67 ŚB: 284–285.

68 AKB VI.10.

69 "*caturthadhyānavivekajaṃ hy ākāśānantyāyatanaṁ, tadvivekajaṁ vijñānānantyāyatanaṁ, tad-vivekajam ākiñcanyāyatanaṁ, tadvivekajaṁ naiva-saṁjñānāsaṁjñāyatanam ity evaṁ catvāra ārūpyāḥ. ko'yaṃ viveko nāma? yena mārgeṇādhastād vimucyate. vairāgyagamanāt*" (AKB VIII.2).

70 "*ye tu bhūtabhāvinaḥ kṣaṇāḥ te pariṇāmānvitāḥ vyākhyeyāḥ*" (YB III.52).

71 To explain how meditation-based knowledge (*bhāvanāmayī prajñā*) liberates, Vasubandhu refers to Vaibhāṣikas who uses the metaphor of a person who is well trained in swimming and therefore is capable of swimming independently up to the other shore. ("*suśikṣitaḥ plavan nirapekṣas tarati*" (AKB VI.5cd)).

4

ON ISOLATION (*KAIVALYAPĀDAḤ*)

In the final chapter of *Yogasūtra*, consisting of thirty-four aphorisms, called *Kaivalyapādaḥ*, Patañjali deals with the journey of the yogin towards final emancipation, that is, *kaivalya*, as termed in Sāṅkhya. He also engages in philosophical debate with the Buddhist doctrines of momentariness and idealism. Here is an aphorism-wise summary of the third chapter of the *Yogasūtra*.

(1–6:) Patañjali begins the chapter by resuming his discussion of supernormal powers. He recognizes birth, drugs, spells (*mantra*), penance and concentration as the sources of supernormal powers. Then he deals with "transformation into another life-state" or "creation of mind" as a power. He describes how a creator mind directs the activities of the created minds. **(7–11:)** He then talks about the law of karma and says that the process of karma and its fruition is beginningless but the law of fruition is inoperative in the case of the actions of a yogin as they are morally neutral. **(12–14:)** He enters then in an argument with Sautrāntika Buddhists who don't believe in continuation of things beyond the present. Patañjali believes that things have continuity in past and future as their identity is determined by their being transformations of the strands of the primordial nature. **(15–24:)** He then has an argument with the idealist Buddhists (*cittamātratāvādins*) who claim that an object is a creation of a mind. Patañjali's position is that of a Sāṅkhya realist, according to whom the same object can be known by many minds which undergo transformation according to the nature of the object. He agrees that the mind is influenced by latent impressions. But the mind according to him is not independent of an external factor. It operates for the sake of the spirit (*puruṣa*), which does not undergo any transformation. **(25–34:)** Patañjali then describes the final spiritual journey of the person who has emancipating knowledge. Since such a person has realized the reality in its particularity, he stops cultivating the feeling of selfhood. His mind then becomes inclined towards discrimination and culminates into isolation. Though some defects may become manifest due to latent impressions even in this state, they can be abandoned through means such as meditative practice. Patañjali then refers to the highest meditative trance called *dharmamegha* (i.e. the virtue-showering cloud), which arises if one does not become lethargic even after attaining the highest knowledge. In this stage, one becomes free from all defilements as well as from the knowledge-obscuring veil. Patañjali concludes by describing the situation in Sāṅkhya terms. Accordingly, in the final stage, the strands of the primordial nature stop producing manifestations as the purpose of the former is fulfilled whereas the spirit exists in its original state.

The comments on these aphorisms will reveal how the Buddhist background consisting of the theory of supernormal powers, the doctrines of momentariness, idealism and the Mahāyāna doctrine of ten planes (*bhūmi*) are relevant for understanding Patañjali's claims.

जन्मौषधिमन्त्रतपः समाधिजाः सिद्धयः ।।1।।

Janmauṣadhimantratapaḥsamādhijāḥ siddhayḥ.

Supernormal achievements come from birth, drugs, spells (*mantra*), penance or concentration.

Comments

Here Patañjali is giving a classification of supernormal achievements (*siddhi*) in terms of their possible origins. Patañjali's classification tallies with Vasubandhu's classification of supernormal powers (*ṛddhi*). It appears that Patañjali has simply used the term *siddhi* by replacing the Buddhist term *ṛddhi*. Vasubandhu says

> In brief *ṛddhi* is described as of five kinds: result of meditation, attained by birth, arising from chants, generated by drugs and caused by actions.[1]

Its closeness to Patañjali's classification is given in Table 4.1.

जात्यन्तरपरिणामः प्रकृत्यापूरात् ।।2।।

Jātyantarapariṇāmaḥ prakṛtyāpūrāt.

The transformation into another life-state results from filling up of (one's own) nature in it.

TABLE 4.1 *Siddhis* (Yoga) and *ṛddhis* (Buddhism)

Sl. no.	*Classification of* siddhis *according to* Yogasūtra	*Classification of* ṛddhis *according to Vasubandhu*
1	*janma (-ja)*	*upapattija/upapattilābhika*
2	*auṣadhi (-ja)*	*auṣadha (ja)*
3	*mantra (-ja)*	*Mantraja*
4	*tapaḥ (-ja)*	*Karmaja*
5	*samādhi (-ja)*	*bhāvanājā/bhāvanāphala*

Sources: YS IV.1, AK and AKB VII.53

Comments

In this and the next three aphorisms Patañjali is trying to give explanation of the creation of minds. *Jātyantara* means a different birth. In this context it is generation of a different mind.

In *Abhidharmakośa* (AK VII.48ab), Vasubandhu talks about two forms or expressions of supernormal powers (*ṛddhis*) as (1) movement (*gati*) and (2) creation (*nirmāṇa*).

Under movement as a supernormal power, Vasubandhu included movement of body according to one's resolve and mind-like speedy movement, which we have considered before while discussing YS III.48.

Under creation (*nirmāṇa*), Vasubandhu discusses creation of things as well as creation of minds (*nirmāṇacitta*).[2]

Here we can use the terminology of creator mind and created mind (*nirmātṛ-citta* and *nirmāṇacitta*). Creator mind can be called primary mind and created mind as secondary mind. According to Patañjali, generation of a new mind by a yogin is a kind of *pariṇāma* (transformation) made possible by filling up (*āpūra*) one's own nature (*prakṛti*) in it. Hence Patañjali is explaining "creation of mind" by using Sāṅkhya terminology.

निमित्तमप्रयोजकं प्रकृतीनां वरणभेदस्तु ततः क्षेत्रिकवत् ।।3।।

Nimittam aprayojakaṃ prakṛtīnāṃ varaṇabhedas tu tataḥ kṣetrikavat.

The causal conditions (such as merit and demerit) are non-determiners of the natures (of the created minds). The difference in their veils, however, is similar to (the case of) a farmer (who cultivates different plants differently).

Comments

The theme of "creation of minds" (which really means creation of living beings) continues. Patañjali, like Vasubandhu, wants to say that the secondary mind does not have its own personality. The natures of the secondary mind or minds are determined by the primary mind. Now if there are many secondary minds, they might have different natures. In the case of primary minds, their nature is determined by their own merit and demerit as their causal conditions (*nimitta*). But in the case of secondary minds, Patañjali argues, such causal conditions are non-determiners (*aprayojaka*).

The word *varaṇabheda* in the aphorism can be interpreted in at least three ways:

(1) Difference in choice (*varaṇa*)
(2) Difference in covering (veil) (*varaṇa* = *āvaraṇa*)
(3) Breaking the cover (*bheda* = breaking) (*varaṇa* = *āvaraṇa*)

Vyāsa accepts the third meaning and gives a complex explanation. It hardly explains the phenomenon of diversity among created minds.

I prefer the second meaning. According to it, *varaṇabheda* means *āvaraṇabheda*, that is, difference in veils. As Vasubandhu explains (AKB IV.96), every living being is characterized by three veils: veil of karma (*karmāvaraṇa*), veil of maturation (*vipākāvaraṇa*) and veil of defilements (*kleśāvaraṇa*). But the thickness and density of these veils may differ from being to being. How to explain this difference is a question. Patañjali is saying that the diversity among secondary minds is caused by the primary mind. This can be explained by the example of *kṣetrika* (a farmer); Patañjali seems to mean that just as a farmer who cultivates different plants differently is responsible for the differential growth of the plants, similarly the creator mind which creates different secondary minds is the cause of their differential status.

निर्माणचित्तान्यस्मितामात्रात् ।।4।।

Nirmāṇacittāny asmitāmātrāt.

All created minds are constructed from ego alone.

Comments

Here Patañjali uses the word *nirmāṇacitta* explicitly, which was implied in earlier aphorisms. He is explaining here how the "created minds" arise. Probably this has reference to Vasubandhu's explanation of the same theme. According to Vasubandhu, there are fourteen types of created minds which arise from the four types of meditation. The general rule he follows is that from a given type of meditation, the mind belonging to the same sphere and lower spheres can be created. Table 4.2 explains the creation of fourteen minds.

Patañjali is also talking about "many created minds", since he is using the word *nirmāṇacitta* in plural. But for him those minds are due to *asmitā* only ("*asmitāmātrāt*"). Here prima facie the positions of Vasubandhu and Patañjali fall apart.

There is perhaps a way out. According to Patañjali's classification of "*samprajñāta-samādhi*", the first stage of *samādhi* has four elements: *vitarka*, *vicāra*, *ānanda* and

TABLE 4.2 Creation of minds through meditation

Meditation (dhyāna) *no.*	*The minds created* (nirmāṇacitta)
1	One mind of sensuous world + one mind belonging to first *dhyāna* = 2
2	The above 2 + the mind belonging to second *dhyāna* = 3
3	The above 3 + the mind belonging to the third *dhyāna* = 4
4	The above 4 + the mind belonging to the fourth *dhyāna* = 5
Total	2+3 + 4+5=14

Source: AKB VII.50ab

asmitā. The earlier elements drop out one after the other in the later stages. In the fourth stage, only the last element, namely *asmitā*, remains.[3] This implies that *asmitā* is the common factor of all the stages of *samprañāta-samādhi*. Hence, the expression "*asmitāmātrāt*" can be taken to mean that the created mind arises "only from the I-notion", which is common to all the stages of *samprañāta-samādhi*.

Hence, Patañjali's account comes somewhat close to that of Vasubandhu, because, as we have seen, Vasubandhu also maintains that all the four *dhyāna*s can lead to "created minds".[4] This could be the connecting link between Vasubandhu and Patañjali.

प्रवृत्तिभेदे प्रयोजकं चित्तमेकमनेकेषाम् ।।5।।
Pravṛttibhede prayojakaṁ cittam ekam anekeṣām.
The one mind (that is, the creator mind) directs the different activities of many minds.

Comments

Patañjali here is referring to one mind (creator mind) stimulating many created minds in different activities. This idea seems comparable to Vasubandhu's reference to the simultaneous speech activity by many created minds. Vasubandhu refers to many created minds speaking along with the creator mind: "When many minds are created, they speak simultaneously. When one mind speaks, the created minds speak with it. When one mind keeps mum all (the created ones) keep mum".[5]

तत्र ध्यानजमनाशयम् ।।6।।
Tatra dhyānajam anāśayam.
Of those, the meditation-born is free from impressions (*āśaya*).

Comments

Here Patañjali is saying that the created mind (*nirmāṇacitta*) which arises from meditation is without latent impressions. In this sense, it is neither good nor bad in itself. It is neutral. This is comparable to Vasubandhu's statement in *Abhidharmakośa*, viz. "the mind arising from meditation is neither-good-nor-bad (*avyākṛta*)". Vasubandhu explains: The mind caused by birth (*upapattija*), for example, that of gods, snakes or ghosts, can be either good (*kuśala*) or bad (*akuśala*) or neutral (*avyākṛta*). But the mind caused by meditation is only neutral.[6]

कर्माशुक्लाकृष्णं योगिनस्त्रिविधमितरेषाम् ।।7।।
Karmāśuklākṛṣṇaṁ yoginas trividham itareṣām.
An action of a yogin is neither white nor black (that is, neither meritorious nor de-meritorious); that of others is either of the three kinds (viz. white, black and both white and black).

Comments

The discussion of creator mind and created mind has come to conclusion in the last aphorism. From this aphorism onward, five aphorisms are devoted to the question of operation of the doctrine of karma.

In continuation with the idea expressed in the last aphorism about the good, bad and neutral mind, Patañjali here is referring to a four-fold action, in terms of its effect. Ordinary persons perform actions with (benevolent or harmful) defilements and hence they yield unpleasant or pleasant fruit. This results in bondage. A yogin performs actions without defilement and hence his actions do not cause bondage. The action performed with benevolent defilement can be called "white" (*śukla*), the action performed with harmful defilement could be called "black" (*kṛṣṇa*) and the action performed with mixed defilement can be called "both white and black". The action performed without defilement can be called neither white nor black (*aśuklākṛṣṇa*). This four-fold classification of actions introduced by Patañjali literally follows the Buddhist classification. As Vasubandhu, in his commentary on AK IV.59cd, says,

> There exists the black action which has black maturation (*vipāka*). There exists the white action which has white maturation. There exists the action both-black-and-white which has both-black-and-white maturation. And there exists an action which is neither black nor white which is without maturation. This kind of action leads to destruction of actions.[7]

The Buddhist classification of actions in terms of their colours might have been influenced by the Jaina classification of actions in terms of six colours and the corresponding classification of souls (*jīva*) in terms of their colours (famously called *leśyā*). According to Jainas the black colour of *jīva* represents extreme immorality, the white colour extreme piety and other colours intermediate degrees of immorality and morality. Moreover, the souls in bondage have all the six colours, whereas the finally liberated soul which has no association with actions is colourless.[8]

In spite of the points of similarity between the Jaina and the Buddhist account, one major point of difference also needs to be noted. Jainas believed that karma is made of matter (*pudgala*) and hence it actually has different colours. Similarly, karma physically sticks to the *jīva* by which the *jīva* actually becomes coloured or looks coloured. When, however, Buddhists classify karma in terms of colours, they seem to use colour-words in metaphorical sense, and not in literal sense.

ततस्तद्विपाकानुगुणानामेवाभिव्यक्तिर्वासनानाम् ॥8॥

Tatas tadvipākānuguṇānām evābhivyaktir vāsanānām.

Thence are manifested the subconscious impressions appropriate to their maturation.

Comments

In the last aphorism Patañjali classified actions into good, bad, mixed and neutral. This classification is based on fruition or maturation (*vipāka*) of actions. Experience of pleasure or pain is fruit of an action. Here Patañjali is bringing out the relationship between action and its fruit through the notions of maturation (*vipāka*) and latent impression (*vāsanā*). Actions generate latent impressions which result in the fruits. There is a problem here. A person may have performed many actions of different kinds, but the impressions of all of them do not yield fruits simultaneously. Only those impressions yield fruits which are appropriate for maturation.

These ideas have a background in Buddhism.[9] Vasubandhu brings out the relationship between action and pleasant/painful/neutral sensations through maturation by saying that an action is experienced as pleasant or painful or neutral through maturation.[10] But how does this maturation take place? Karma (an action) itself does not get matured. Asaṅga explains in *Yogācārabhūmi*:[11]

> An action is performed under the influence of some impressions due to which it becomes good or bad. The action arises and ceases. But due to that, as conditioned by (the goodness or badness) it produces a specific series of impressions (*saṁskāra*). This series of impressions is called *vāsanā*. The *vāsanā*-series finally culminates into favorable or unfavorable fruit.

Patañjali's statement here reflects this Asaṅga-Vasubandhu theory of *vāsanā*.

जातिदेशकालव्यवहितानामप्यानन्तर्यं स्मृतिसंस्कारयोरेकरूपत्वात् ।।9।।

Jātideśakālavyavahitānām apy ānantaryaṁ smṛtisaṁskārayor ekarūpatvāt.

They (that is, the consequent experiences) are (to be regarded as) immediate though they are distanced by birth, space and time, because memory and impressions are identical in nature.

Comments

Patañjali, in this aphorism, is explaining the interrelation between actions and their results in terms of the Sāṅkhya theory of causation.

An action is the cause of a pleasant/unpleasant sensation. But we generally see that there is a gap between action and the consequent sensation in terms of birth, space and time. An action is performed in one birth, place and time and its consequence is experienced in another birth, place and time. Patañjali is claiming that in spite of such a gap, the consequences are to be treated as immediate.

Patañjali gives the reason that recollection and impression are identical. Impression (*saṁskāra*) is the cause of which recollection is the effect. Here Patañjali is using the word *saṁskāra* with double meaning. In the case of recollection, *saṁskāra* is an impression generated by an experience, which becomes awakened and causes recollection. In the context of karma theory, *saṁskāra* is the latent impression generated by an action which gets matured and causes fruition. Patañjali seems to be claiming that in both the cases, there is identity between cause and effect. This is how Patañjali seems to be employing the Sāṅkhya theory of causation. According to Sāṅkhya an effect pre-exists in its cause, such that there is no gap between cause and effect, but there is identity between them.

तासामनादित्वं चाशिषो नित्यत्वात् ।।10।।

Tāsām anāditvaṃ cāśiṣo nityatvāt.

They (that is, the latent impressions) have no beginning, as the hope (that is, the desire to live) is eternal.

Comments

The word *tāsāṁ* which means "of them" refers to "*vāsanānām*" (that is, "of the latent impressions") from aphorism YS IV.8. Hence Patañjali is saying here that the latent impressions have no beginning. The reason he gives is that the desire to live is permanent.

Patañjali here seems to be referring to the beginningless cycle of our worldly existence. This particular idea does not seem to stem from Sāṅkhya, as the classical Sāṅkhya accepts the origination of the world from the contact between primordial nature and spirits; it does not support the idea of beginningless-ness. Buddhist (and Jaina) tradition, on the other hand, does possess this idea. Vasubandhu, for instance, refers to the beginningless wheel of transmigration in the following words:

> In this way the wheel of becoming (*bhavacakra*) is beginningless.[12]
>
> In this way the birth is caused by defilements and actions and they in their turn are caused by birth and the birth occurs again from them. This is how the wheel of becoming should be known as beginningless.[13]

हेतुफलाश्रयालम्बनैः संगृहीतत्वादेषामभावे तदभावः ।।11।।

Hetuphalāśrayālambanaiḥ saṅgṛhītatvād eṣām abhāve tadabhāvaḥ.

Being accomplished by (main cause) cause, (intended) result, substratum and support (or object), when they (=the causal factors) are not there, it (=the fruition) is not there.

Comments

Patañjali seems to be saying here that human actions yield their results because they are covered (*saṅgṛhīta*) by four conditions: *hetu* (the main cause/efficient cause), *phala* (intended result), *āśraya* (basis, material cause) and *ālambana* (object). Hence, if they are not there, the result of actions will also be not there. This reminds us of the Buddhist four-fold classification of conditions (*pratyaya*s). As Vasubandhu refers to the *sūtra* which gives this classification:

> Four types of conditionalities (*pratyayatā*) are stated. Conditionality of the efficient cause (*hetu*), that of the immediately preceding condition (*samanantara*), that of the object (*ālambana*) and that of the governing factor (*adhipati*).[14]

There is a close correlation (though not complete correlation) between Patañjali's classification and the Buddhist classification, as given in Table 4.3.

The comparison shows that *hetu* and *ālambana* are common to both of the classifications. *Samanantara-prayaya* in the Buddhist classification refers to the immediately preceding item in the causal series. This causal condition is important for Buddhists because they do not accept *ātman* or any substance which is generally accepted as material cause by other systems. Naturally Patañjali, who accepts Sāṅkhya metaphysics, accepts substance (in this case, *prakṛti*) as the material cause or the substantial basis (*āśraya*). This could be the explanation of *āśraya* replacing *samanantara*.

There does not seem to be any direct correlation between *adhipati-pratyaya* accepted by the Buddhists and *phala* as the condition accepted by Patañjali. But there seems to be an indirect correlation, because *adhipati-pratyaya* is understood as the governing cause, the cause which governs the fruit (*phala*).

Hence one can argue that there is at least 75 per cent similarity between the two classifications such that Patañjali's classification is arguably a reconstruction of the Buddhist classification. Patañjali's statement, "When they (=the causal factors) are not there, it (=the fruition) is not there", echoes the negative aspect of the Buddhist formula of dependent origination.

अतीतानागतं स्वरूपतोऽस्त्यध्वभेदाद्धर्माणाम् ।।12।।

Atītānāgataṁ svarūpato'sty adhvabhedād dharmāṇām.

Past and future exist in reality as the phenomena belongs to different dimensions (of time).

Comments

Patañjali is claiming here that the past and future state of a thing are as real as its present state. The past, present and future according to him can be regarded as different temporal dimensions (*adhva-bheda*) of the same thing.

TABLE 4.3 Classification of causes according to Buddhism and Patañjali

Buddhist classification	*Patañjali's classification*
Hetu (efficient cause)	*Hetu* (efficient cause)
Adhipati (governing condition)	*Phala* (intended result)
Samanantara (immediately preceding condition)	*Āśraya* (ground, material cause)
Ālambana (object as cause)	*Ālambana* (object as cause)

Sources: AKB II.61c, YS IV.11

Patañjali's position here seems to be influenced by the Vaibhāṣika/Sarvāstivāda school of Buddhism as described by Vasubandhu in AKB V.25. As Vasubandhu states, "Vaibhāṣikas say that past and future do exist",[15] and "Sarvāstivādins (those who claim that everything exists) are so called because they claim their existence (i.e. the existence of all the three times). They are of four kinds".[16]

Vasubandhu in this context states four positions of Sarvāstivādins. We have discussed the four positions under YS III.13. There we saw that Vyāsa has no difficulty in accepting the first three positions. Here he once again explains them in terms of manifestation. According to him the present thing has the present aspect in manifest form; but this does not mean that it does not have past and future aspects. Only they are non-manifest. The future thing is that which will become manifest later. The past thing is that whose manifestation is already experienced. The present thing is that which is undergoing its own function.[17]

The word *adhvan* means "path". But in the Buddhist literature the word is used in the sense of temporal dimension. Patañjali also has used the word *adhvan* here in the same sense. Now, there is no issue about the existence of things in the present. But whether things exist also in past and future is an issue. On this, Patañjali says,

> The past and the future object exist in reality as well, because the things have different temporal dimensions (*adhvabhedāt dharmāṇām*).

It is possible that here Patañjali has used the word *dharma*, typically in the Buddhist sense "thing/phenomenon" as, for example, Vasubandhu uses the word *dharma* in the sentence—"*yadā dharmaḥ kāritraṁ na karoti tadā'nāgataḥ, yadā karoti tadā pratyutpannaḥ yadā kṛtvā niruddhas tadā'tīta iti*" (AKB V.26) (When a thing/phenomenon does not perform its activity, it is yet to come (i.e. future), when it performs, it is present, when it ceases after performing, it is past). In this way "*adhvabhedāt dharmāṇām*" means "things differ according to their temporal dimensions".

But Vyāsa seems to have avoided the Buddhist connotation, knowingly or unknowingly, and interpreted *dharma* as property. As he says, "*dharmī cāneka-dharmasvabhāvas tasya cādhvabhedena dharmāḥ pratyavasthitāḥ*" (A property-bearer by its nature has many properties. Its properties are established in terms of temporal difference).

Vasubandhu, while explaining the Vaibhāṣika defence of past and future, states four reasons given by Vaibhāṣikas.

> There is existence of all the three times, (1) because it has been said by the Buddha; (2) because cognition arises from two conditions (namely faculties and objects coming together); (3) because an object (of cognition) must be real and (4) because (one performs actions for) future results.
>
> *(AKB V.25b)*

Vyāsa in his commentary gives the last two reasons: (1) an object of cognition must exist and (2) if future does not exist then the actions performed for future results will be meaningless.[18]

Patañjali accepts all the three times because this view is consistent with Sāṅkhya metaphysics. According to the Sāṅkhya theory of causation known as *Satkāryavāda*, there is an inseparable link between past, present and future states of things, because causation is only manifestation.

ते व्यक्तसूक्ष्मा गुणात्मानः ।।13।।

Te vyaktasūkṣmā guṇātmānaḥ.

They, whether they are manifest or subtle, are of the nature of strands (of *prakṛti*).

Comments

Here by the pronoun *te* ("they") Patañjali is referring to *dharmas* (things, phenomena) of the preceding aphorism. Here Patañjali is simply saying that whether the things are manifest or subtle (that is, non-manifest), they are of the nature of the three strands, namely *sattva, rajas* and *tamas*. In a way, this aphorism and the next form the preface for Patañjali's forthcoming criticism of the Buddhist idealism. Here Patañjali is suggesting that manifest or non-manifest status of a thing is not subject-dependent, but objective, determined by the strands of *prakṛti*.

परिणामैकत्वाद्वस्तुतत्त्वम् ।।14।।

Pariṇāmaikatvād vastutattvam.

The identity of a thing is due to sameness of the transformation.

Comments

Here Patañjali is explaining "identity of a thing" (*vastu-tattvam*: "that-ness" of a thing) from the Sāṅkhya point of view. He is saying that identity of a thing is determined by sameness of the transformation (of the three strands of *prakṛti*). In continuation with the preceding aphorism, he implies that the identity of a thing does not depend upon the knower.

Vyāsa makes the criticism of the Buddhist idealism explicit. He refers to idealist Buddhists (without identifying them as Buddhists) as

> those who conceal the nature of a thing by following the line of thought that the object is not associated with consciousness, but consciousness is associated with an object, and say that a thing is only a mental construction like a dream-object; it does not exist in the real sense.[19]

Then he proceeds,

> How can their speech be reliable, as they deny the thing, which is presented as "that" (*tatheti*) by its own worth, by discarding the nature of

the thing on the strength of mental construction which in itself is not an authentic cognition?[20]

Vyāsa here is probably referring to Vasubandhu's argument in *Viṁśatikā* (verses 1 and 3) which compares cognitions with dream and regards objects as unreal.[21]

वस्तुसाम्ये चित्तभेदात्तयोर्विभक्तः पन्थाः ।।15।।

Vastusāmye cittabhedāt tayor vibhaktaḥ panthāḥ.

In spite of the sameness of objects, the minds are different. Hence they follow different paths.

Comments

The anti-idealistic argument continues.

If an object can be inseparably related to a mind, then it would not be cognized by another mind. But we find that the same object is cognized by different minds and it produces different cognitions in different minds. So Patañjali concludes that the object and the mind follow separate courses of existence.

Vyāsa in his commentary refers to an idealist argument: "Some say: The object arises only with the cognition, because it is a consumable object, like pleasure etc.". This was probably the argument of idealist Buddhist, because according to them the feelings such as pleasure do not exist without being cognized. Again this could be an argument advanced against Sāṅkhyas, because according to Sāṅkhyas, pleasure, pain and delusion (*sukha-duḥkha-moha*) are the three strands of *prakṛti* and they exist independently of consciousness.

However, the source of the argument referred to by Vyāsa is not known.

न चैकचित्ततन्त्रं वस्तु तदप्रमाणकं तदा किं स्यात् ।।16।।

Na caikacittatantraṁ vastu tad apramāṇakaṁ tadā kiṃ syāt.

Nor is an object dependent on a single mind; if it were so, then it would not be proved (to be objectively real). What would it be then?

Comments

In this aphorism, Patañjali is questioning the idealist Buddhist explanation of knowledge of things. In the next two aphorisms, he will give the explanation of knowledge from the realist (Sāṅkhya) point of view. Here Patañjali is claiming that the idealist Buddhist view that a thing depends on a single mind for its own existence and knowledge is not correct. For, in that case, there will be no (objective) evidence for the thing.

तदुपरागापेक्षित्वाच्चित्तस्य वस्तु ज्ञाताज्ञातम् ।।17।।

Taduparāgāpekṣitvāc cittasya vastu jñātājñātam.

An object would be (sometimes/partly) known and (sometimes/partly) unknown because the mind depends upon being coloured by the object.

Comments

Here and in the next aphorism, Patañjali is distinguishing between the knowledge of a thing and that of a mental state. A mind, in order to produce knowledge of a thing, requires to be coloured by the object (*taduparāga*), that is, undergo transformation according to the object. Hence, an object can be known, subject to transformation of the mind or simply remain unknown if such a transformation does not take place. So, the same object may be sometimes cognized, sometimes not. Or among many objects, some are cognized, some are not. A mental state, on the other hand, cannot remain unknown. It is always present to some *puruṣa*. *Puruṣa* is aware of a mental state without undergoing any transformation. Patañjali will state this in the next aphorism.

सदा ज्ञाताश्चित्तवृत्तयस्तत्प्रभोः पुरुषस्यापरिणामित्वात् ।।18।।

Sadā jñātāś cittavṛttayas tatprabhoḥ puruṣasyāpariṇāmitvāt.

To its lord, the self (*puruṣa*), the modifications of the mind are always known, as he is changeless.

Comments

Here Patañjali is saying that although transformation in mind (*citta*) is needed for the knowledge of an object, the transformation in mind itself is known to *puruṣa* without any transformation in the latter. *Puruṣa* is aware of every transformation in mind (or state of mind—*citta-vṛtti*) whenever it takes place, without being coloured by the transformation. Hence the duality of objects and minds on the one hand—which undergo transformation—and *puruṣa* on the other hand—which does not undergo transformation—is maintained.

न तत्स्वाभासं दृश्यत्वात् ।।19।।

Na tat svābhāsaṁ dṛśyatvāt.

The mind does not appear to itself, as it is the object of seeing (and not the subject).

Comments

Now Patañjali addresses another Buddhist doctrine, namely the self-manifesting character of consciousness. This doctrine is not restricted to idealist Buddhists. Self-manifesting character of consciousness was accepted by Sautrāntikas also. In fact the doctrine of self-manifestation has at least two versions: idealist and realist.

(1) **Idealist version:** This can be called an eliminative version. According to it when the mind cognizes an object, in fact it cognizes itself (because there is no object over and above consciousness).

(2) **Realist version:** This can be called a non-eliminative version. Accordingly, the mind which cognizes an object also cognizes itself at the same time (through the same cognition).

Patañjali in this aphorism and the next two aphorisms is criticizing the idealist version of the doctrine of self-manifestation.[22]

According to Patañjali, the mind is not self-manifesting (*svābhāsam*) because it is only an object (of *puruṣa*). According to Sāṅkhya only *puruṣa* is *draṣṭā* and everything else (mind as well as objects of mind) is *dṛśya*.

एकसमये चोभयानवधारणम् ।।20।।

Ekasamaye cobhayānavadhāraṇam.

Both (i.e. itself and something else) cannot be cognized simultaneously.

Comments

Those who hold a self-manifesting nature of mind assimilate mind with lamp or fire. A lamp or fire illuminates an object but it cannot illuminate an object without illuminating itself.

As against this, Patañjali is arguing that a mind cannot perform two cognitive acts (cognizing itself and cognizing an object) simultaneously.

Vyāsa explains that the believer of momentariness (*kṣaṇikavādin*, that is, a Buddhist) maintains that whatever is the being of a thing, the same is its activity and the same is the producer.[23] According to this doctrine, the Buddhist maintains that consciousness can be both object-conscious and self-conscious simultaneously. Patañjali is questioning this view of the Buddhist.

चित्तान्तरदृश्ये बुद्धिबुद्धेरतिप्रसङ्गः स्मृतिसङ्करश्च ।।21।।

Cittāntaradṛśye buddhibuddher atiprasaṅgaḥ smṛtisaṅkaraś ca.

If it were to be known by another mind, there would be an undesirable consequence (*atiprasaṅga*) with reference to "cognition of cognition" and admixture with recollection.

Comments

Patañjali continues his argument. In this aphorism he is probably stating the possible objection which the Buddhist (who regards cognition as self-manifesting) can raise. The objection runs as follows. If a cognition is cognized not by that cognition itself but by another cognition (that is, by immediately succeeding cognition—*samanantara* as Vyāsa explains), there will be two difficulties. (1) There will be undesirable consequence (*atiprasaṅga*) with reference to cognition of cognition and (2) admixture with recollection. The difficulties may be explained as follows:

(1) **Undesirable consequence:** If every cognition is cognized by another cognition, then the cognition of cognition will have to be cognized by some third cognition and so on, ad infinitum.

(2) **Admixture with recollection:** If the next cognition cognizes the previous cognition, then it will be regarded as memory cognition. Hence, the cognition of cognition (that is, the next cognition) will have to be regarded as a direct cognition as well as a memory cognition simultaneously.

चितेरप्रतिसंक्रमायास्तदाकारापत्तौ स्वबुद्धिसंवेदनम् ।।22।।

Citer apratisaṅkramāyās tadākārāpattau svabuddhisaṁvedanam.

The self, which is un-transferable, cognizes the cognition which belongs to it by assuming likeness with the form of cognition.

Comments

In this and the next two aphorisms, Patañjali is giving a Sāṅkhya-type explanation of knowledge. Through these aphorisms, he, as a Sāṅkhya metaphysician, is trying to answer the idealist Buddhist by saying that (1) the cognition (*buddhi*) is not cognized by itself, or by another cognition, but by *puruṣa*, and (2) *buddhi* does not operate for itself but for the sake of *puruṣa*. Out of these two statements, the first statement is made in this aphorism. *Citi* in the aphorism refers to *puruṣa*. *Puruṣa* has awareness of *buddhi*. *Buddhi* is the *sva* (property, own thing) of *puruṣa*, hence, it is called *svabuddhi*. *Puruṣa* is described here as *apratisaṅkrama*, that is, without transference or transition. But it as if assumes the form of "that" (here "that" should mean the mind or *buddhi*) and hence has the sensation/awareness of the *buddhi*.

Here the phrase "as if", which is not in the aphorism, is used by the commentator. Perhaps the word *āpatti* in the aphorism may have that suggestion. *Puruṣa* by its nature is *apariṇāmī*. It does not undergo any transformation. But when knowledge takes place, it "as if" becomes like the *buddhi*. The implication of this for the present context is that unlike the idealist Buddhist who would say that a cognition is cognized by itself, a Sāṅkhyaite would say that a cognition is not cognized by itself, but by *puruṣa*, to which the cognition is said to belong.

द्रष्टृदृश्योपरक्तं चित्तं सर्वार्थम् ।।23।।

Draṣṭṛdṛśyoparaktaṁ cittaṁ sarvārtham.

The mind coloured by the knower and the known is all-comprehensive.

Comments

Here Patañjali seems to be claiming that mind can have everything as its object. It can have "seer/observer" (*draṣṭā*) as well as "seeable/observable" (*dṛśya*) as its object. When the mind has anything as its object, the former gets "coloured" (*uparakta*) by the latter. It assumes the forms of the latter. It imitates the latter.

Vyāsa adds here a comment against the idealist school of Buddhism. He says:

> This capacity of mind (like that of a crystal) of reflecting everything has misled some who regard the mind itself to be conscious. There are others

who think that everything is mind only ("*cittamātramevedaṁ sarvam*") and that this world including (the sentient beings namely) cows and others and (the insentient things namely) pots and others along with their causes does not exist. Those (idealists) are pitiable. Why? For they still possess the seed of illusion that mind itself has appearances of every form.

Vyāsa then argues that in meditative insight one cannot reduce an object to consciousness and that one will have to accept *puruṣa* as the knower.

From the side of the idealist Buddhists, it could be argued that the concept of *puruṣa*, who is supposed to know everything without undergoing any change, is problematic.

तदसंख्येयवासनाभिश्चित्रमपि परार्थं संहत्यकारित्वात् ।।24।।

Tad asaṅkhyeyavāsanābhiś citram api parārthaṁ saṃhatyakāritvāt.

Though it (the mind) is complex with innumerable subconscious tendencies (*vāsanā*), it exists for the sake of another, as it acts as a combination.

Comments

Here Patañjali is giving a typical Sāṅkhya argument for the existence of *puruṣa*. According to the Sāṅkhya argument, everything which is a combination of things (*saṅghāta*) is meant for something other than itself. For example, bed, chair, etc. are not meant for themselves but for something else, namely, the person who uses them. Similarly, these complex things such as sense-organs, mind and body are meant not for themselves but for something other than them, and that other thing is the *puruṣa*.

That the mind is a complex of latent impressions is the view of Yogācāra Buddhists. In fact, this is their concept of *ālayavijñāna*. As Chatterjee (1971: 18) comments, "it is so termed as it is the place or receptum, the storehouse containing the seeds or impressions (*vāsanā*) of all future experiences".

Patañjali seems to agree with this. His use of the word "*api*" ("even though") suggests this. He, however, does not accept the implications that the Buddhists draw from it. For Buddhists, the mind out of its latent impressions and passions is active for its own pleasure and removal of its own suffering. Patañjali, on the other hand, tries to establish the existence of *puruṣa* by using the teleological argument of Sāṅkhya.

We find that in later times Diṅnāga criticized this Sāṅkhya argument as a case of the fallacy called *iṣṭavighātakṛt* (the fallacious reason property which proves the opposite of the intended target property).[24] According to him, the Sāṅkhyas intend to prove with this argument, the existence of *puruṣa*, which is a non-composite (simple) object. But because the things such as beds and chairs are meant for concrete persons which are composite entities by their very nature, the argument, instead of proving the existence of the non-composite *puruṣa*, will prove some composite entity, which is opposed to the intended target property.

विशेषदर्शिन आत्मभावभावनाविनिवृत्तिः ।।25।।

Viśeṣadarśina ātmabhāvabhāvanāvinivṛttiḥ.

One who has realized the particularity, stops cultivating the feeling of selfhood.

Comments

The argument against the idealist Buddhist was over in the last aphorism. From this aphorism onwards Patañjali describes the yogin's journey to the final liberation. This description is partly influenced by the Buddhist account, as Patañjali derives some elements of it from Mahāyāna description of *bodhisattva*'s journey towards Buddha-hood (as we shall see in aphorism IV.29).

In the present aphorism, Patañjali seems to refer to the stage of *ṛtambharā prajñā* (truth bearing wisdom) (YS I.49), which Patañjali describes as direct insight (in the sense that it is different from scriptural or inferential cognition) because it has "particular" as its object (*viśeṣārthatvāt*). So, the seer of particularity (*viśeṣadarśī*) who is mentioned here seems to be the yogin who has the truth-bearing wisdom. We have seen there that this truth-bearing wisdom is close to *bhāvanāmayī prajñā* of Buddhism. But here one more Buddhist element is added and that is the cessation of the feeling of "selfhood" (*ātmabhāva-bhāvanā-vinivṛtti*). This idea seems to be influenced by the no-self theory of Buddhism. It seems that Patañjali's approach oscillates between the Brahmanical doctrine of self and the no-self doctrine of Buddhism. For example, in YS II.42 he refers to "eligibility for perceiving *ātman*" (*ātmadarśanayogyatva*) as a result of mental purification. This seems to be a result of Brahmanical influence. But the aphorism YS II.5 (which regards accepting self in place of no-self as a form of misconception) as well as the present aphorism indicates the influence of the no-self theory of Buddhism.

तदा विवेकनिम्नं कैवल्यप्राग्भारञ्चित्तम् ।।26।।

Tadā vivekanimnaṁ kaivalyaprāgbhārañ cittam.

At that time mind is inclined towards discrimination and culminates into isolation (*kaivalya*).

Comments

This is a reconstruction of a Buddhist description of an emancipated person (*arhat*), which Vasubandhu quotes from *Aṅgārakarṣūpama-sūtra*:

> *vivekanimnaṁ cittaṁ bhavati yāvan nirvāṇaprāgbhāram.*
>
> *(AKB VI.58, pp. 376–377)*

(The mind (of the emancipated person) is inclined towards discrimination and so on finally culminating[25] into emancipation (*nirvāṇa*)). Patañjali seems to have borrowed the statement by replacing the word *nirvāṇa* (the Buddhist term for emancipation) by *kaivalya* (the Sāṅkhya term for emancipation).

Similarly, although Patañjali adopts the term *viveka-nimnam* from the Buddhist quotation, here there is probably a shift in the meaning of the term *viveka* also. The shift is from moral-psychological connotation of discrimination[26] accepted in Buddhism to the metaphysical sense of discrimination accepted in Sāṅkhya.

तच्छिद्रेषु प्रत्ययान्तराणि संस्कारेभ्यः ।।27।।

Tacchidreṣu pratyayāntarāṇi saṁskārebhyaḥ.

In their intervals there arise other experiences due to the (past) latent impressions.

Comments

Here Patañjali is describing the penultimate state of a yogin in which certain deficiencies can become manifest due to the remainder of the latent impressions. This is comparable with the possibility of spiritual degeneration of an *arhat*, who has still not achieved the final liberation according to Abhidharma Buddhism.

In *Abhidharmakośabhāṣya* (AKB VI.56–58), Vasubandhu conducts a long discussion in which Vaibhāṣikas talk about six types of *arhat*. Out of them, the sixth type, namely *akopyadharmā* (one who cannot be shaken), is the highest type of *arhat* who cannot regress. The other five types are slightly inferior. They attain a free state (*vimukti*) only subject to certain conditions (*samaya*). These five are deficient either because they in their student stage did not practice meditation with continuity (*sātatya*) or devotion (*satkṛtya*) or with neither. Or, though they practiced meditation with continuity and devotion, their faculties are weak (*mṛdvindriya*).[27] As against this Sautrāntikas maintain that no regress is possible from *arhat*-hood.[28]

Who was right in this controversy—Vaibhāṣikas or Sautrāntikas—is not the issue here. The point is that Patañjali here is concerned with the same kind of situation with which Vasubandhu was concerned—that of the possibility of regress in the penultimate stage of emancipation. Patañjali says that the person who has attained realization may have gaps (*chidra*)[29] in his continuous flow of meditative experience due to impressions of the past karmas. Such a person has other experiences (*pratyayāntarāṇi*) (that is, probably mundane experiences) which mark his imperfection.

हानमेषां क्लेशवदुक्तम् ।।28।।

Hānam eṣāṃ kleśavad uktam.

Their abandonment is stated like that of defilements.

Comments

The theme of the preceding aphorism, namely the signs of imperfection which can arise in the person who has attained realization, continues. Now the question is how to overcome the signs of imperfection and attain perfection.

In Vasubandhu's terminology, this is the case of *aśaikṣaparipūrṇatva*. *Aśaikṣa* is the one who has crossed learner's stage—as he has realized the truth, and *paripūrṇatva* means accomplishment, perfection. Vasubandhu states that this is possible by using two methods:[30] (1) by sharpening/strengthening the faculties (*indriya*) and (2) by attaining meditative absorption (*samāpatti*). We have seen that according to Vaibhāṣikas the deficiencies in emancipation are either because one has not practiced meditation with continuity and devotion or because one's faculties are weak (*mṛdvindriya*). Hence, one has to remove these deficiencies to attain perfection.

Patañjali's answer is similar. After all, the imperfection is due to latent impressions. In YS II.10–11, Patañjali had said that defilements are abandoned either by reverse creation or meditation. Here Patañjali is saying that the same methods are to be employed at the penultimate stage also.

प्रसंख्यानेऽप्यकुसीदस्य सर्वथा विवेकख्यातेर्धर्ममेघः समाधिः ।।29।।

Prasaṅkhyāne'py akusīdasya sarvathā vivekakhyāter dharmameghaḥ samādhiḥ.

One who does not become lethargic even after attaining the highest knowledge attains discriminating knowledge in every respect, which leads to the meditative trance (known as) *dharmamegha* (i.e. the virtue-showering cloud).

Comments

Here Patañjali is referring to the highest state of meditative absorption, which he calls *dharmamegha-samādhi*. Neither Patañjali nor Vyāsa give any explanation as to why the highest *samādhi* is so called. Patañjali seems to have borrowed this term from the Mahāyāna Buddhist theory of ten planes (*daśa-bhūmi*) of the *bodhisattva*'s spiritual progress. The highest plane is called *dharmameghā*[31] *bhūmi* there. This indicates Patañjali's tendency to appropriate different Buddhist spiritual achievements and show them as the achievements possible in the framework of Sāṅkhya-based Yoga. In Buddhism the doctrine of ten *bhūmi*s presupposes the doctrine of *pāramitā*s (perfections). Every *bhūmi* is a result of some spiritual perfection. Patañjali here is only talking about *dharma-megha* and he is treating it as the result of "complete discriminating knowledge" (*sarvathā viveka-khyāti*), which is the liberating knowledge according to Sāṅkhya.

It is quite probable that Patañjali has Asaṅga's *Bodhisattvabhūmi* as a background. In *Bodhisattvabhūmi* Asaṅga discusses different stages of *bodhisattva*'s spiritual development (*vihāras*) and correlates them with the ten planes (*bhūmis*) as described in *Daśabhūmīśvara-sūtra*.[32] He correlates the highest stage of *bodhisattva* with *dharmameghā bhūmi*. Asaṅga brings out the significance of the term *dharmameghā* in the following words:

> The highest stage of spiritual life (*paramavihāra*) should be understood elaborately in accordance with the *sūtra* as it is described in the *bodhisattva*-plane called "that of the cloud of virtue" as follows:
>
> The *bodhisattva*, who has completed the course of *bodhisattva* and fully accumulated the wealth of enlightenment, receives from the Buddhas the

> shower of the right virtue, which is the cloud of virtue, which is extremely large and therefore unbearable by other beings. Then he himself becomes the cloud of virtue and when his enlightenment is not awakened or when it is awakened, he pacifies the dust of defilement of innumerable beings by his shower of the right virtue. When he stays in that plane, he sows diverse grains of wholesome roots, cultivates them and makes them fully mature. That is why this plane is called as that of the cloud of virtue and is called the highest spiritual stage in the same sense.[33]

Patañjali says that the yogin attains *dharmamegha-samādhi* when he becomes *akusīda*, even after having emancipating knowledge (*jñānāvaraṇīya*). The meaning of the term *akusīda* can be understood only on the Buddhist background. Vyāsa has not taken the Buddhist significance into account and has interpreted the word as "desireless and detached",[34] which is baseless.

Buddhist Abhidharma uses the word *kausīdya* (abstract noun from *kusīda*) to mean lethargy, lack of enthusiasm, a quality which is opposite of *vīrya* (energy, enthusiasm).[35]

Patañjali's contention seems to be that after attaining insight into truth (*jñānāvaraṇīya*) one is likely to be complacent and lethargic. Due to that there would occur gaps (*chidra*) in his spiritual progress. So even after the insight the yogin should continue his efforts and fill up the gaps in his spiritual character—so that he achieves complete knowledge (*sarvathā vivekakhyāti*).

The term *prasaṁkhyāna* which Patañjali uses in the sense of insight into truth, or spiritual knowledge, gives an impression that he is talking about Sāṅkhya type of spiritual knowledge. But this need not be the case. Just as the word *prasṁkhyāna* has affinity with the word Sāṅkhya (because *sam* +√*khyā* is common to the roots of both), it has also affinity with the terms *pratisaṁkhyā*, *pratisaṁkhyāna* and *samprakhyāna* which refer to knowledge or realization of the noble truths according to Buddhism. Vasubandhu defines *pratisaṁkhyā* as knowing or realising (*pratisṁkhyānam*) the noble truths such as suffering. He calls it a special type of wisdom (*prajñāviśeṣa*)[36]. Similarly, he uses the term *samprakhyāna* in the sense of wisdom (*prajñā*) while explaining misconception (*avidyā*) as the absence of *samprakhyāna*.[37] Hence it is possible that Patañjali's use of the term *prasaṁkhyāna* is influenced by the Buddhist conception of spiritual knowledge expressed by the terms *pratisaṁkhyā* and *samprakhyāna*.

ततः क्लेशकर्मनिवृत्तिः ॥30॥

Tataḥ kleśakarmanivṛttiḥ.

Then there is stoppage of defilements and actions.

Comments

Patañjali in his presentation of Yoga system presents the aspects of the system at a moral-psychological level as well as a metaphysical level. The moral-psychological level is largely influenced by Buddhism, whereas the metaphysical level is influenced by Sāṅkhya. Presently, the theme is the final journey of the liberated person. What happens in that stage at the moral-psychological level is stated in this and the next aphorism. What happens at the metaphysical level is stated in aphorism IV.32.

Patañjali is saying here that due to *dharmamegha-samādhi* there is stoppage of defilements and actions. By "stoppage of actions" he probably means the stoppage of defiled actions (which produce latent impressions, that is, *vāsanās*). This seems to be the reflection of Asaṅga's description of the *bodhisattva*'s highest spiritual stage (*paramavihāra*). Asaṅga says that in the highest stage there is abandonment of all defilements, latent impressions and dormant defilements as well.[38]

तदा सर्वावरणमलापेतस्य ज्ञानस्यानन्त्याज्ज्ञेयमल्पम् ।।31।।

Tadā sarvāvaraṇamalāpetasya jñānasyānantyāj jñeyam alpam.

Then because of infiniteness of knowledge from which all obscuring impurities have been removed, little remains to be known.

Comments

Here Patañjali is saying that in the liberated state, all the knowledge-obscuring impurities are destroyed and hence the yogin becomes almost omniscient.

According to Mahāyāna Buddhism there are two types of veils: the veil of defilement (*kleśāvaraṇa*) and the veil of knowable (*jñeyāvaraṇa*). As we referred to Asaṅga's statement, in the final stage of the *bodhisattva*, the veil of defilement gets destroyed. But the veil of knowable does not get destroyed in the *bodhisattva* stage. It gets destroyed completely only in the Buddha stage, which Asaṅga calls *tāthāgata-vihāra*. While describing the extreme purity of the Buddha's knowledge of all objects, Asaṅga says:

> What is the (complete) purity of knowledge? When the badness (*dauṣṭhulya*) belonging to all ignorance is removed, the knowledge becomes uncovered with regard to all knowable objects and one obtains mastery over knowledge. This is called complete purity of knowledge.[39]

Patañjali seems to be having this idea at the background. In other words, Patañjali is indirectly referring to the Buddha stage which immediately follows the highest stage of the *bodhisattva*.

The idea is also close to Jainism, according to which the *jīva* has intrinsic capacity for infinite knowledge (*ananta-jñāna*), but it is obscured by the kārmic inflows of eight kinds,[40] including "knowledge covering karma" (*jñānāvaraṇīya karma*). According to Jainas, when the *jīva* attains liberation, the knowledge-obscuring karmas are destroyed and the *jīva* is restored to its original nature, which is characterized by omniscience.

Though both Buddhist and Jaina contexts are relevant to this aphorism of Patañjali, the Buddhist context seems to be more relevant because of the reference to "*dharmamegha-samādhi*" in YS IV.29.

Vyāsa tries to explain the idea in Sāṅkhya terms as mind, which is pure *sattva*, becomes free from the veil of *tamas* and becomes operative occasionally by *rajas*. It is doubtful whether Patañjali has all this in mind.

ततः कृतार्थानां परिणामक्रमसमाप्तिर्गुणानाम् ।।32।।

Tataḥ kṛtārthānāṃ pariṇāmakramasamāptir guṇānām.

Thence there is an end to the sequence of the modifications of the strands (of *prakṛti*) as they have fulfilled their purpose.

Comments

Here Patañjali is explaining the final stage of emancipation in Sāṅkhya terminology. In fact, the last three aphorisms reflected the Buddhist description (probably Asaṅga's description) of the *bodhisattva*'s final journey from *bodhisattva*-hood to Buddha-hood. One would expect that Patañjali would continue the same story. But he changes the track and shifts to the Sāṅkhya explanation of the final journey of a spiritual practitioner, probably because he wants to give prominence to Sāṅkhya.

Accordingly, in the final stage of the spiritual journey of a practitioner, the process (*krama*: an ordered sequence) of transformation of the strands of *prakṛti* comes to an end, because the strands have fulfilled their purpose (that is, *puruṣārtha*—namely *mokṣa*).

क्षणप्रतियोगी परिणामापरान्तनिर्ग्राह्यः क्रमः ।।33।।

Kṣaṇapratiyogī pariṇāmāparāntanirgrāhyaḥ kramaḥ.

The sequential order is apprehensible in terms of the transformation and the utmost limit; it is opposed to momentariness.

Comments

Here Patañjali is talking about the sequential order (or an ordered sequence) with reference to objects which undergo transformation. He is saying that an ordered sequence can be specified in terms of two aspects: transformation (*pariṇāma*) and its utmost limit (*aparānta*).[41] Applying this concept of *krama* to *guṇas* (that is, the strands of *prakṛti*) he wants to point out that the process of transformation belonging to the strands, which is virtually the process of transmigration, is a continuous process, but it comes to an end.

He is also saying here that a sequential order is opposed to momentariness (*kṣaṇapratiyogī*). By this he implies that the two features, namely "transformation" and "utmost limit", cannot be found in a momentary object. In this way, Patañjali is also trying to criticize the Buddhist doctrine of momentariness through this aphorism.

In his commentary, Vyāsa distinguishes between two kinds of permanence: permanence of an object undergoing transformation (*pariṇāminityatā*) and absolute permanence (*kūṭasthanityatā*). He also claims that sequential order is found in every permanent thing. He applies the notion of sequential order to both of the types of permanent things. It is hard to know why he wants to apply the notion to *puruṣa* also, when Patañjali applies it only to *guṇas*.

Vyāsa makes another point however which clearly exhibits Buddhist influence. He asks whether *saṁsāra* (the world/the cycle of transmigration) comes to an end or not and says that this should not be said (*avacanīyam etat*). He refers here to the classification of questions into *ekānta-vacanīya* (which can be answered categorically), *vibhajya-vacanīya* (which can be answered after dividing them, splitting them into alternatives) and *avacanīya* (which cannot be answered). This classification reflects the classification which Vasubandhu refers to in AKB V.22. Vyāsa, while giving the classification of questions, replaces the term *vyākaraṇīya* with *vacanīya*. The difference in terminology is given in Table 4.4.

By using this classification, Vyāsa concludes that the question whether *saṁsāra* has an end or not has no definite answer. So the question is *avacanīya*. This question and this answer resemble the question "Is *loka sānta* or *ananta* or both or neither?" and the Buddha's silence over it.[42] But Vyāsa also says that the question whether *saṁsāra* has an end or not is an answerable (*vyākaraṇīya*) question. He also tries to answer the question by saying that *saṁsāra* (transmigration) of a virtuous (*kuśala*) person comes to an end, that of others doesn't. One might say here that Vyāsa is expressing his difference with the Buddha, because whereas the question for the Buddha is unanswerable, it can be answered (after splitting it into two alternatives) according to Vyāsa. Such a conclusion would be hasty

TABLE 4.4 Types of questions according to Buddhism and Vyāsa

Buddhist terminology	*Vyāsas terminology*
Ekāṁśavyākaraṇīya	*Ekānta-vacanīya*
Vibhajyavyākaraṇīya	*Vibhajya-vacanīya*
Paripṛcchya-vyākaraṇīya	*Paripṛṣṭe vibhajya-vacanīya*
Sthāpanīya	*Avacanīya*

Sources: AKB V.22, YB IV.33

because the question which was not answered by the Buddha was about the world (*loka*), whereas the question which Vyāsa was answering was about transmigration (*saṁsāra*).

Though Vyāsa's classification of questions is influenced by the Buddhist classification, whether Vyāsa is adopting the Buddhist classification consistently can be doubted. For example, the Buddhist classification includes a variety called "*paripṛcchya-vyākaraṇīa praśna*", which means the question which can be answered after asking questions for clarification. This variety is not clearly available in Vyāsa's version. Whether he refers to it by the nomenclature "*paripṛṣṭe vibhajya-vacanīya*" is unclear. Similarly, he calls the last type of question as *avacanīya* (unsayable) and also *vyākaraṇīya* (answerable). This, too, is unclear.

पुरुषार्थशून्यानां गुणानां प्रतिप्रसवः कैवल्यं स्वरूपप्रतिष्ठा वा चितिशक्तिरिति ।।34।।

Puruṣārthaśūnyānāṃ guṇānāṁ pratiprasavaḥ kaivalyaṁ svarūpapratiṣṭhā vā citiśaktir iti.

Isolation (*kaivalya*) is the reverse generation of the strands (of *prakṛti*) when they don't have (to fulfil) any purpose of the self or it means the power of consciousness (i.e. *Puruṣa*) established in its own nature.

Comments

Patañjali concludes his work by giving a Sāṅkhya-type metaphysical description of the final emancipation (*kaivalya*). It is the reverse creation[43] of the strands of *prakṛti*, which are empty of the so-called goals of *puruṣa*.[44] To describe the same state from the side of the power of consciousness (*puruṣa*): the power of consciousness simply remains in its original state.

Notes

1 This is Vasubandhu's explanation of the verse AK VII.53, which runs as follows:

> *avyākṛtaṁ bhāvanājaṁ trividhaṁ tūpapattijam / ṛddhir mantrauṣadhābhyāṁ ca karmajā ceti pañcadhā //*
>
> Result of meditation is neutral (neither wholesome nor unwholesome). The mind produced by birth is of three kinds (wholesome, unwholesome and neither). *Ṛddhi* (supernormal power) is of five kinds—(That caused by meditation, by birth,) that caused by *mantra*, by drugs and by action.

2 According to Vasubandhu, things are not created directly by supernormal powers. First, a mind is created and through the created mind, an object is created. "*kiṁ khalv abhijñayaiva nirmāṇaṁ nirmīyate? Nety ucyate. kiṁ tarhi? abhijñāphlaiḥ nirmāṇacittaiḥ*" (AK and AKB VII.49).

3 See my comments on YS I.17. I have shown there that Patañjali's theory of *samprajñāta-samādhi* (as interpreted by Vyāsa) follows the same policy of deletion as the Buddhist theory of four *dhyānas* does.

4 However, an important difference remains. In Patañjali's scheme, *asmitā* plays a double role: as one of the defilements (equivalent to *asmimāna* of Buddhism) and as a factor in meditation. This second role Patañjali assigns to *asmitā* is rather puzzling. Vyāsa also seems to have realised this problem. See my comments in this connection under YS I.17.

5 *yadā bahavo nirmitā bhavanti, tadā yugapad bhāṣante:*
ekasya bhāṣamāṇasya bhāṣante saha nirmitāḥ /
ekasya tūṣṇīṁbhūtasya sarve tūṣṇīṁ bhavanti te //

(*AKB VII.51*)

6 "*avyākṛtaṁ bhāvanājaṁ, trividhaṁ tūpapattijaṁ/*" "*upapattipratilambhikaṁ tu nirmāṇacittaṁ kuśalākuślam avyākṛtaṁ bhavati devanāgapiśācādīnām*" (AK and AKB VII.53ab).

7 "*asti karma kṛṣṇaṁ kṛṣṇavipākam. asti karma śuklaṁ śuklavipākam. asti karma kṛṣṇaśuklaṁ kṛṣṇaśuklavipākam. asti karma akṛṣṇam aśuklam avipākam. yat tat karma karmakṣayāya saṁvartate iti*" (AKB IV.59cd). Asaṅga presents this classification in ADS: 59. Tandon (1998: 4) refers to similar passage from Anguttaranikāya 4.24.2.1.

8 Thanks to Maithrimurthi for drawing my attention to the Jaina theory of karma in this context. For the Jaina theory of the colours of *jīvas* caused by their association with karma, see Glasenapp (1942: 47–49, 75–92).

9 The influence of the Buddhist doctrine of *vāsanā* on the Yoga school was also acknowledged by Frauwallner (1984: 328–329).

10 "*vipākavedanīyatā karmaṇaḥ*" (AKB IV.49).

11 "*yeṣu saṁskāreṣu yac chubhāśubhaṁ karma utpannaniruddhaṁ bhavati, tena hetunā tena pratyayena viśiṣṭā saṁskārasantatiḥ pravartate sā vāsanā ity ucyate, yasyāḥ prabandhapatitāyā iṣṭāniṣṭaphalaṁ nirvartate*" (YAB: 127–28).

12 "*ity anādi bhavacakrakam*" (AK VIII.19).

13 "*anena prakāreṇa kleśakarmahetukam janma, tadhetukāni punaḥ kleśakarmāṇi tebhyaḥ punarjanma ity anādi bhavacakrakaṁ veditavyam*" (AKB III.19).

14 "*sūtre catasraḥ pratyayatāḥ. hetupratyayatā, ālambanapratyayatā samanantarapratyayatā, adhipatipratyayatā ceti*" (AKB II.61).

15 "*asty evātītānāgatam iti vaibhāṣikāḥ*" (AKB V.25b). Asaṅga, too, refers to this position, which he calls "*atītānāgatadravyasadbhāva*" ("The existence of past and future substances") as one of the rival positions in *Yogācārabhūmi* (YAB: 122). The present aphorism seems to be a response to Asaṅga's criticism of the position.

16 "*tadastivādāt sarvāstivādā iṣṭāḥ caturvidhāḥ*" (AK V.25cd).

17 "*bhaviṣyadvyaktikam anāgatam. anubhūtavyaktikam atītam. svavyāpāropārūḍhaṁ vartamānam*" (YB IV.12).

18 Vasubandhu states the two reasons as "*sadviṣayāt phalāt*" (AK V.25b).

19 "*nasty artho vijñānasahacaraḥ, asti tu jñānam arthasahacaraṁ svapnādau kalpitam, anayā diśā ye vastusvarūpam apahnuvate; jñānaparikalpanāmātraṁ vastu, svapnaviṣayopamaṁ tu na paramārthato' stīti ye āhuḥ*" (YB IV.14).

20 "*te tatheti pratyupasthitam idaṁ svamāhātmyena vastu katham apramāṇātmakena vikalpajñānabalena vastusvarūpam utsṛjya tad evāpalapantaḥ śraddheyavacanāḥ syuḥ?*" (YB IV.14).

21 "*vijñaptimātram evaitad asadarthāvabhāsanāt*" (Vi 1ab) [All this is only consciousness, because there is appearance of unreal objects.] "*deśādiniyamaḥ siddhaḥ svapnavat*" (Vi 3ab) [The rule that things appear in specific space(, time,) etc. (does not prove that things are real), because in a dream also things are proved to be like that.]

22 According to *Pātañjala-yoga* edited by Brahmalīnamuni, YS IV.19 is preceded by Vyāsa's commentary in which Vyāsa refers to those who are "*vaināśika cittātmavādin*" as the

opponents to whom Patañjali is responding. The expression "*vaināśika cittātmavādin*" obviously refers to the idealist Buddhists. Here the term "*cittātmavādin*" should be interpreted as "the one who holds that everything is of the nature of mind". In this sense "*cittātmavādin*" means "*cittamātratāvādin*".

23 "*kṣaṇikavādino yad bhavanaṁ saiva kriyā tadeva ca kārakam ity abhyupagamaḥ*" (YB IV.20).

24 For my discussion of the fallacy, see Gokhale (1992: 104) and Gokhale (2016: 104–105).

25 According Edgerton's Buddhist Hybrid Sanskrit Dictionary, the words *nimna* and *prāgbhāra* are almost synonymous, meaning "inclined to" or "bent upon" or "headed for". But according to Apte's Dictionary, *prāgbhāra* means top or summit of a mountain. Maithrimurthi brought to my notice that Monier-Williams gives the meaning "the slope of a mountain" according to which "inclining/bending" makes more sense.

26 In Buddhism, the term *viveka* is used in the sense of separation and is classified into three kinds: bodily seclusion, mental purification and final liberation (*kāya-viveka, cittaviveka* and *avadhi-viveka*). Davids and Stede, Pali English Dictionary: "*viveka*").

27 "*akopyadharmā yo naiva parihātuṁ bhavyaḥ. prathamau dvau pūrvam eva śaikṣāvasthāyām sātatyasatkṛtyaprayogavikalau. tṛtīyaḥ sātayaprayogī, caturthaḥ satkṛtyaprayogī pañcama ubhayathāprayogī mṛdvindriyas tu. ṣaṣṭha ubhayathāprayogī tīkṇendriyaś ca*" (AKB VI.57cd).

28 "*arhattvādapi nāsti parihāṇir iti sautrāntikāḥ*" (AKB VI.58b (p. 375)).

29 The term *chidra* in the sense of gaps or holes in spiritual practice is found used by Asaṅga in his description of the stages (*vihāra*) in spiritual development. (For Asaṅga's theory of *vihāras* see Appendix I, Part II). According to him, a yogin acts narrowly and with gaps up to the second stage ("*parīttakārī bhavati sacchidrakārī*", BSB: p. 220). From the third stage onwards, he acts immeasurably and without gaps ("*apramāṇakārī bhavati acchidrakārī*", BSB: 220).

30 "*aśaikṣaparipūrṇatvaṁ dvābhyām*" (AK VI.65a).

31 The word *dharmameghā* is the adjective of *bhūmi* and hence is in feminine gender. Patañjali uses it as the adjective of *samādhi* and hence in masculine gender. The compound word *dharmamegha* can be split as *Bahuvrīhi* compound such that the meaning would be, "That in which *dharma* is (like) a cloud".

32 For details, see Appendix I, Part III.

33 "*paramo vihāro veditavyaḥ vistaranirdeśataḥ punar yathāsūtram eva tad yathā dharmameghāyāṁ bodhisattvabhūmau. paripūrṇabodhisattvamārgaḥ suparipūrṇabodhisambhāraś ca sa bodhisattvaḥ tathāgatānām antikād dharmameghabhūtām atyudārāṁ dussahāṁ tadanyaiḥ sarvasattvaiḥ saddharmavṛṣṭiṁ sampratīcchati.mahāmeghabhūtaś ca svayam anabhisambuddhabodhir abhisambuddhabodhiś cāprameyānāṁ sattvānāṁ saddharmavṛṣṭyā nirupamayā kleśarajāṁsi praśamayati. vicitrāṇi ca kuśalamūlasasyāni virohayati vivardhayati pācayati tasyāṁ bhūmau avasthitaḥ. tasmāt sā bhūmir dharmameghety ucyate. tenaiva cārthena paramo vihāro draṣṭavyaḥ*" (BSB: 242). For other details, see Appendix I, Part III.

34 "*na kiñcit prārthayate, tatrāpi viraktasya*" (YB IV.29).

35 Vasubandhu includes *kauśīdya* in the list of *kleśa-mahābhūmikadharmas* (the factors which always occur in defiled mind) and defines it as "*cetaso nābhyutsāho vīryavipakṣaḥ*" (non-enthusiasm of mind, the opposite of energy). See AK and AKB II.26ab.

36 "*duḥkhādīnām āryasatyānāṁ pratisaṁkhyānaṁ pratisaṁkhyā prajñāviśeṣaḥ*", AKB I.6ab

37 For Vasubandhu's use of the terms *samprakhyāna* and *asamprakhyāna*, see AKB III.29d

38 "*parame punar vihāre sarvakleśavāsanānuśayāvaraṇaprahāṇaṁ veditavyam*" (BSB: 243).

39 " *tatra katamājñānaviśuddhiḥ? pūrvavat sarvāvidyāpakṣyadauṣṭhulyāpagamāt sarvatra ca jñeye jñānasyānāvaraṇāt jñānavaśavartiā sarvākārā jñānaviśuddhir ity ucyate*" (BSB: 265). This statement belongs to the chapter called *Pratiṣṭhā-paṭala* (the last chapter of BSB) which is devoted to the description of the Buddha stage.

40 For eight types of karma, see Tatia (2007: 33–34).

41 The word *aparānta* (rather "*aparānta koṭi*") is used in the case of *saṁsāra* in Buddhism. According to Buddhism, *saṁsāra* has an extreme limit (end). See the entry "*aparānta-koṭi*" in Edgerton's Buddhist Hybrid Sanskrit Dictionary.

42 For the list of unanswered questions in Buddhism, see Jayatilleke (2010: 243).

43 For the notion of *pratiprasava* (reverse creation), see the comments on YS II.10–11.

44 For the role of the concept of *puruṣārtha*, see the comments on YS II.21–22.

5

CONCLUDING OBSERVATIONS

The interpretation of the Yoga aphorisms and the comments on them offered in the last four chapters hopefully indicate that many terms, concepts and doctrines of Patañjali's Yoga can be traced to Buddhism, particularly the Abhidharma Buddhism of Vasubandhu and the early Yogācāra of Asaṅga. Consideration of this Buddhist background can lead us to a better and proper understanding of Patañjali's Yoga.

Two stories of suffering and emancipation

The reconsideration of Patañjali's yoga of the above sort presents a different picture of his theory of yoga. According to it, Patañjali through his theory of Yoga tries to synthesize two stories of suffering and emancipation which he develops through his work side by side or in a mixed way. One story can be called empirical-cum-psychological, one which seems to be rooted largely in Buddhism. It seems that Patañjali must have had grounding in Buddhist Abhidharma literature and early Yogācāra literature, and, perhaps, he must have practiced different stages in the Buddhist meditation. All this seems to be reflected in his empirical-cum-psychological story of suffering and emancipation. His studies in Buddhism must have been supplemented by his acquaintance with Jainism also.

The second story of suffering and emancipation Patañjali develops is rooted in his studies and belief in the metaphysics of Sāṅkhya. Probably the Sāṅkhya philosophy which Patañjali had before him was not exactly the Sāṅkhya of *Ṣaṣṭitantra*, which was later on crystalized by Īśvarakṛṣṇa in *Sāṅkhyakārikā*, but rather the version of Sāṅkhya influenced by Vindhyavāsin.[1]

It is possible to sketch out the two stories by referring to the aphorisms of the *Yogasūtra* themselves and then try to understand what Patañjali seems to have done with the two stories. (While sketching out the stories the *Yogasūtra* aphorism numbers will be given in brackets.)

I. The empirical–psychological story of suffering and emancipation

In order to attain the final cessation of suffering one has to focus on four foundational concepts: suffering which is to be abandoned, the cause of what is to be abandoned, the actual abandonment and the means by which what is to be abandoned is abandoned (II.16–17, 25–26). For a discriminating person, everything is an object of suffering. Certain things result in suffering, certain things are painful by their very nature, and others are objects of suffering due to their conditioned character (II.15).

We experience suffering because of our defilements (*kleśa*). They are five in number, namely misconception (*avidyā*), egoism, attachment, hatred and adherence to dogmatic views (II.3). Misconception consists in apprehending the things which are non-eternal, impure, painful and non-self as eternal, pure, pleasant and self-containing respectively (II.5). Misconception is the breeding ground of all other defilements (II.4). These defilements cause pleasant and painful experiences either in this birth or the next, through maturation of actions (II.12–14). The actions which ordinary people perform contain defilements and hence they are either "good" or "bad" (white or black) or mixtures of goodness or badness. Hence, they lead to fruition through latent impressions. Yogin's actions are neither white nor black. Hence, they do not lead to fruition (IV.7).

The state of mind free from suffering is naturally the state free from misconception and other defilements, and also free from the latent impressions of the past actions. The latent impressions can be called *vāsanā*, *āśaya* or *bīja* (seed). They have no beginning though they can come to an end (IV.8–11). Hence the highest state of mind is devoid of the seed of latent impressions. It is the state of seedless meditative absorption. It is also the state of the cessation of all mental states (I.51).

Such a highest state can be achieved by the path of self-discipline, which mainly consists of two factors: practice (of cultivation of virtues and meditation: *abhyāsa*) and detachment (*vairāgya*) (I.12). The practice should be continued for a long time, without interruption and with devotion (I.14). Detachment is of two kinds: ordinary and higher. Initially, one develops detachment towards pleasant, sensuous objects, but ultimately one develops the higher detachment which is towards all objects (I.15–16).

The path of purification of the mind begins with following the rules of conduct such as non-injury, truthfulness, non-stealing, celibacy and non-possession (II.30). These rules of conduct are to be followed universally and unconditionally, as a great vow (II.31).

One has also to develop physical and mental cleanliness, contentment and tolerance towards opposite pairs. One has to study spiritual texts and make a resolve to become an ideal being (II.32). By sitting in a stable and comfortable posture one should observe vigilantly the incoming and outgoing breath outside, inside and in the middle column in terms of location, duration and number and also experience the state in which breathing stops temporarily (II.46–51). One should practice withdrawal of the senses from the objects and make the senses inward-directed (II.54).

Thereafter one should practice meditation. Some beings have an inborn capacity for attaining a meditative trance whereas others have to use certain means. There are five such means: faith, energy, mindfulness, concentration and wisdom (I.19–20). Meditative practice develops through the stages of fixed attention, homogenous free flow of the mind and absorption of the mind in the object0 (III.1–3). In the initial stages of meditative absorption (*samādhi*), one entertains gross objects of thought. Gradually they drop out, and one entertains subtle objects of thought. Gradually they also drop out and one experiences bliss (I.17). At a higher stage, one entertains objectless meditative absorption, with the remainder of latent impressions (I.18). Through the practice of this type of meditation one has direct insight (*prajñā*) into the reality, which is different from both wisdom based on scriptural knowledge and wisdom based on reasoning (I.48–49). This insight helps in the reduction of the latent impressions. When one becomes free from latent impressions, one attains a seedless meditative trance in which there is cessation of all mental states (I.51).

When one masters all the stages of meditation and integrates them (III.4), one develops a special insight useful for the achievement of supernormal knowledges and powers (III.5–6). Hence one can know one's own previous births (III.18). One can know the mental states of others, though one does not know the objects of those mental states (III.19–20). One can disappear (III.21). One can attain power comparable to that of an elephant (III.24). One can have extrasensory perception (III.36); for example, one acquires a divine ear (III.41). One can walk through thorns or mud without their sticking to him (III.39). One can travel through the sky (III.42). One achieves a very beautiful and strong body (III.46). One can travel with the speed of mind (III.48). One can become extremely small or large or light or heavy (III.45). One can become omniscient (III.49). An advanced yogin can even create new minds and control their activities (IV.2–5). But though such supernormal powers are achievements of extravert life, they are obstacles on the path to meditative absorption (III.37).

Even after attaining the knowledge of the truth, one should not stop the efforts and be lethargic. Thus, one attains the highest plane of *samādhi* which is called the virtue-showering cloud (*dharmamegha*) (IV.29).

II. The metaphysical story of suffering and emancipation

The metaphysical story also holds that everything is an object of suffering. But the reason it gives is that everything is a combination of the three strands (*guṇa*) of *prakṛti* and that the three strands contained in each thing are in a state of imbalance and mutual conflict (II.15).

According to this story also, suffering arises from *avidyā* (II.24). But this *avidyā* is *aviveka*, that is non-discrimination between the "seer" (*draṣṭā*, *puruṣa*) and the objective world which is "seen" (*dṛśya*). Suffering arises when the seer and the seen become united because of non-discrimination (II.17).

Though the most important distinction Sāṅkhya makes is between the strands (*guṇas*) of primordial nature (*prakṛti*) on the one hand and the spirit (*puruṣa*) on the other, Yoga suggests a subtler distinction between two different powers: seer power (*dṛkśakti*) and seeing power (*darśanaśakti*). When these two powers get as if united, I-notion (*asmitā*) arises (II.6). Out of the three strands, namely *sattva*, *rajas* and *tamas*, *sattva*, being the most luminous one, resembles the spirit; hence, it gets confused with the spirit. Due to this confusion, arise pleasant and painful experiences (*bhoga*) (III.35).

The mind (the *sattva*-dominated manifestation of primordial nature) and the spirit play distinct roles in knowledge, bondage, meditative trance and liberation. Knowledge (*pramāṇa*) is one of the modifications of the mind (*citta-vṛtti*) (I.5–6). In the process of knowledge, the spirit plays the role of a passive observer. The spirit, being pure consciousness, does not see the object directly but sees it as reflected in cognition (that is, in a modification of mind) (II.20). The difference between the mind and the spirit is that the mind while cognizing an object undergoes transformation according to the object. The spirit, however, while it is aware of the changing mind, does not undergo any transformation (*apariṇāmī*) (IV.17–18).

In the state of meditative trance, the mind is devoid of any modifications (I.2). At that time the spirit retains its own nature (I.3). Similarly, the state of isolation (*kaivalya*) is marked by the power of consciousness (*citiśakti*, the same as spirit) restoring its original nature (IV.34). In the state other than meditative trance, that is, the state of bondage, however, the spirit gets assimilated with mental modifications (I.4).

The three strands of primordial nature, when in a state of imbalance, create different manifestations and the empirical world comes into existence. The strands, however, do not create manifestations for themselves, but in order to fulfil the goals of a spirit (II.21). When the goals are fulfilled, the strands stop producing manifestations or they withdraw their manifestations. This process is called reverse creation (*pratiprasava*) (IV.34). It is this process of reverse creation through which the subtle defilements are removed (II.10).

Primordial nature is one, whereas the spirits are many. When primordial nature withdraws its manifestations, it does so for the spirit whose goals are fulfilled. It does not do so for other spirits, so that the common empirical world continues to exist for them (II.22).

Though a spirit does not undergo any transformation by itself, when it is in the state of bondage, the transformations in the mind happen to be projected on the spirit. Apart from ordinary spirits, who are in bondage and can become liberated, there is a spirit which is absolutely free from bondage-creating factors such as actions, fruition and latent impressions (I.24). Such a spirit is the ideal spirit and can be given the status of the lord (*īśvara*) (I.24), though it is not the creator

God. This lord possesses the seed of omniscience and, because he is eternal, he plays the role of teacher of the most ancient teachers (I.25–26).

Establishing links between the two stories

Patañjali's discipline of Yoga, in this way, seems to be underlying two stories concerning the four basic truths of life: suffering, the cause thereof, the cessation thereof and the means to the cessation. One story is derived from Abhidharma literature and the early Yogācāra literature of Buddhism (with some inputs from Jainism) and the other story is derived from or based on Sāṅkhya metaphysics. The two stories do not completely merge into one story, though an appearance is created that it is a single story of Yoga (neither that of Buddhism nor that of metaphysical Sāṅkhya). The episodes in the two stories are stated in a mixed way, one after another or sometimes side by side to create an impression that the two stories are not isolated from each other. We also find Patañjali using different ways or tricks for linking the two stories. He gives the metaphysical terminology of Sāṅkhya a phenomenological shape, makes Sāṅkhya categories the objects of meditation, arranges the categories of the two systems hierarchically such that Sāṅkhya categories get prominence and general acceptance at the metaphysical level. Let us see how this happens.

(A) Phenomenological reformulation of Sāṅkhya categories

Yoga system is primarily concerned with empirical-practical-phenomenological matters rather than abstract metaphysical ones. Hence, even after accepting Sāṅkhya metaphysics, Patañjali often presents Sāṅkhya metaphysical categories in the terms closer to experience. Hence, for *prakṛti* he uses the words *guṇas* or *dṛśya*; for *puruṣa* he uses the words *draṣṭā*, *dṛk*, *dṛśi* or *citiśakti* and for *mahat*, *buddhi* or *manas*[2] he uses the words *darśanaśakti* or *citta*. This also brings Sāṅkhya metaphysics closer to Buddhism.

(B) Sāṅkhya categories are made objects of meditation

According to Sāṅkhya the emancipating knowledge of the 25 categories or that of the distinction between primordial nature and the spirit is attained through analogical reasoning (*sāmānyatodṛṣṭa anumāna*) or verbal testimony (*āgama*), because the categories involved in such knowledge are abstract, non-empirical. According to Buddhist soteriology, on the other hand, emancipating knowledge has to be attained through meditative trance, which is non-conceptual, direct. Patañjali tries to overcome this contrast by making Sāṅkhya categories themselves objects of meditation. (1) According to the theory of meditative attainment, gross objects are the objects of *savitarka-samāpatti*, whereas subtle objects are the objects of *savicāra-samāpatti*. Hence, all manifestations of the three strands of primordial nature, gross or subtle, can be objects of meditative attainment according to Patañjali (YS I.44–45). (2) Patañjali brings even the spirit (*puruṣa*), which is totally abstract, within the reach of meditation. According to Sāṅkhya, primordial nature produces manifestations not for itself but for the sake of a spirit. The spirit, on the other hand, is there for its own sake. Hence, Patañjali claims, by practicing integrated concentration on the notion of intrinsicality ("for-itself"), one gets an insight into the spirit (YS III.35). (3) Patañjali postulates "lord" or "supreme being" (*īśvara*) by extending the concept of spirit to its ideal limit. Like the concept of spirit, *īśvara*, too, is a transcendent concept. However, Patañjali suggests that meditation on *īśvara* is possible by repeating the syllable *Om* which he regards as a symbol designating *īśvara*. Because of his ideal nature, meditation on *īśvara* is supposed to have special advantages in terms of removal of impurities and achievement of concentration (YS I.29, II.2).

(C) Sāṅkhya narrative given prominence

While reading the *sūtras* one finds that the episodes of empirical (Buddhist) narrative and metaphysical (Sāṅkhya) narrative are stated one after the other. There, many times the metaphysical narrative seems to become prominent. One feels that according to Patañjali, Sāṅkhya metaphysics sets the framework in which the empirical episodes happen. (1) Patañjali starts with the definition of Yoga as cessation of mental states (YS I.2), which is quite experiential. But in the next aphorism (YS I.3) he refers to the simultaneous status of *draṣṭā* (that is, the spirit), which is transcendent. (2) In the concluding aphorisms (YS IV.25–34) Patañjali mixes the empirical Buddhist terminology (stoppage of the self-notion, abandonment of latent impressions and defilements, the highest meditative trance called "the cloud of goodness" and knowledge becoming free from the veils of impurity) with Sāṅkhya terminology (*viveka, kaivalya, jñānāvaraṇīya, guṇa* etc.), but states the final conclusion in purely Sāṅkhya metaphysical terms (YS IV.34). This again suggests that Patañjali accepted a Sāṅkhya-type metaphysical description of emancipation as ultimate. (3) In the second chapter, Patañjali introduces the Buddhist model of four basic truths—*heya, heyahetu, hāna* and *hānopāya*—but explains all four in Sāṅkhya terms. Hence, the Buddhist narrative of defilements (*kleśa*) as the cause of suffering becomes subordinate. (4) On the question of abandoning defilements, Patañjali says that the gross manifestations of defilements can be abandoned by meditation, but the subtle defilements can be abandoned by reverse creation (*pratiprasava*: a typically Sāṅkhya metaphysical concept). This shows Patañjali's preference for the Sāṅkhya metaphysical narrative at a subtler level.

But Patañjali does not seem to follow this policy consistently. Sometimes a Buddhist narrative attains a superior status in his scheme. Patañjali defines *avidyā*, the most basic defilement in terms of the four perversions (*viparyāsa*) accepted by Buddhists (YS II.4–5). However, he defines *asmitā*, which is a secondary defilement, in Sāṅkhya terms (YS II.6).

(D) Assertion of Sāṅkhya ontology against Buddhist ontology

Bringing the two stories of suffering and emancipation together is not easy because their ontologies contrast with each other. Here Patañjali defends Sāṅkhya ontology against different aspects of Buddhist ontology. (1) Buddhists held that when we say that a thing changes, in fact "properties" (qualia, phenomena, *dharmas*) come into existence and cease to exist; there is no property-bearer over and above properties. This explanation of change is not acceptable to Sāṅkhyas. According to them, properties undergo change when the property-bearer remains intact. Change is nothing but transformation (*pariṇāma*) in which only the form changes, the matter remains the same. This view is called *pariṇāmavāda*. In this controversy, Patañjali asserts the *pariṇāmavāda* model of change which includes the existence of a property-bearer over and above the properties (YS III.14). (2) Another related issue is that of the existence of phenomena in all the three times: the past, the present and the future. Buddhists of the Sarvāstivāda school held that the phenomena (*dharma*) had all the three temporal dimensions but they differed on the question as to how they have them. Vasubandhu refers to four different views (AKB V.25–26). According to the *Bhāvānyathika* view form or structure (*bhāva*) of a thing changes when the substantial feature (*dravyabhāva*) remains common. According to the *Lakṣaṇānyathika* view, a thing has all the three (temporal) dimensions. When a thing possesses any one dimension, the other two dimensions remain latent. According to the *Avasthānyathika* view, temporal dimension of a thing is determined by its activity (*kāritra*). According to the *Anyathānyathika* view, a temporal dimension of a thing is relative to temporal dimensions of other things. After referring to the four positions Vasubandhu comments on them. He accepts the third position, namely *Avasthānyathika*, which explains temporal dimension of a thing in terms of its activity. He rejects the first position because it is nothing but *pariṇāmavāda* of Sāṅkhya. Patañjali responds to this controversy in *Yogasūtra*. Though he does not seem to accept the fourth position, which is relativistic, he has no harm in accepting the first three positions in the framework of *pariṇāma*. Hence, he re-terms the three modalities of change as "*dharmapariṇāma*", "*lakṣaṇapariṇāma*" and "*avasthāpariṇāma*" respectively (YS III.13). (3) Buddhists held that everything that is real undergoes change. There is nothing which remains unchanged. As against this, Sāṅkhyas hold that though primordial nature and its manifestations undergo change (in which they follow the *pariṇāmavāda* model), the spirits remain unchanged. They are *apariṇāmin*. Patañjali accepts this view of Sāṅkhya (YS IV.18). (4) Among the Buddhist schools current at the time of Patañjali, one was idealist, it held that there is only mind but there are no objects external to consciousness. This position was contrary to the Sāṅkhya position according to which primordial nature is objectively real and available simultaneously to all spirits. Here Patañjali defends Sāṅkhya realism against Buddhist idealism.

In this way, although Patañjali adopts so many inputs from the Buddhist story of suffering and emancipation, he tries to situate them in the realist and eternalist Sāṅkhya framework.

Oddities in the attempted synthesis

Of course, we cannot say that Patañjali succeeds in linking up the two stories in a fully consistent way. His attempt to synthesize the two schools gives rise to many odd situations. (1) *Citta*, which is central to Buddhism, is sentient by its very nature. In Patañjali's scheme *citta* replaces *buddhi* or *manas* of Sāṅkhya, which are insentient by their very nature. (2) Following Sāṅkhya, Patañjali describes *puruṣa* as absolutely non-changing, but cannot retain its non-changing nature consistently. He holds that in the extravert state of mind, the seer becomes similar to the mental state (YS I.4). (3) Patañjali's definition of *avidyā* (YS II.5), which is modelled on the Buddhist concept of perversion (*viparyāsa*), causes some problems. The latter presupposes the doctrine of three characters (*trilakṣṇa*), which states that all phenomena are impermanent, soulless and objects of suffering (*anitya*, *anātma* and *duḥkha*). This is partly contrary to Sāṅkhya, which accepts two types of eternality, transformative eternality and absolute eternality, and which also accepts eternal *puruṣa* equivalent to *ātman*. (4) Particularly, Patañjali does not seem to have a clear position on the question of *ātman*. While adopting the Buddhist concept of perversion, he seems to reject the doctrine of self. Similarly, he recommends the realization of particularity, through which one stops cultivating the feeling of selfhood (YS IV.25). But he recommends meditation on the notion of "for-itself" which results in the knowledge of *puruṣa* (YS III.35), and he also praises the capacity for the realization of the self (*ātmadarśanayogyatva*) as a result of (mental) cleanliness (YS II.41). (5) *Asmitā*, equivalent to the I-notion (*ahaṁkāra*) of Sāṅkhya and to self-centric conceit (*asmimāna*) of Buddhism, plays two contradictory roles in Patañjali's scheme. On the one hand, it is a defilement, which is to be abandoned through meditation (YS II.3, II.11). On the other hand, it is the highest object of object-conscious meditative absorption (*samprajñāta-samādhi*) (YS I.17). (6) The spiritual journey of a yogin as described by Patañjali involves leaps from one narrative to the other and again from the other to the first. So, it does not remain a smooth experiential process. It is not clear how from the empirical–psychological experience of the cessation of mental states one is led to the discriminating knowledge (*vivekakhyāti*) of the Sāṅkhya metaphysical type. Or, in other words, how the destruction of mental impurities, brought about by the practice of eight-limbed yoga takes one to Sāṅkhya-type metaphysical knowledge (YS II.28). It is not clear again how from the Sāṅkhya-type metaphysical knowledge (*jñānāvaraṇīya*) one goes to the highest *samādhi* (*dharmamegha-samādhi*) (YS IV.29) of the Mahāyāna Buddhist type, and again, how from that *samādhi* one achieves the final goal, namely *kaivalya*, which is of the Sāṅkhya metaphysical type (YS IV.30–34).

So, what we have in *Yogasūtra* is not a homogenous merger of different streams which have lost their separate identities, but a heterogenous mixture of the streams which sometimes mix up but sometimes collide with each other.

Some significant findings

When I started searching for the possible Buddhist sources of Patañjali's aphorisms and Vyāsa's commentary, the search started throwing new light on many aspects of Pātañjala-yoga. Many concepts and doctrines which appeared unconnected got connected through their Buddhist background. Many loose ideas got tightened; some old ideas had to be discarded. Here are a few novel and significant results.

(1) Patañjali's five-fold classification of defilements (*kleśa*) (YS II.3) seems to be derived from the six-fold classification of *kleśa/anuśaya* given by Asaṅga and Vasubandhu.

(2) Patañjali's division of *kleśa*s into the *pratiprasavaheya* and *dhyānaheya* (YS II.10–11) seems to be the reflection of Vasubandhu's division into *dṛgheya* (to be abandoned through realization) and *bhāvanāheya* (to be abandoned through meditation).

(3) Patañjali considers practice (*abhyāsa*) and detachment (*vairāgya*) as the two means to the cessation of mental modifications (YS I.12). His explanation of these two means reflects analogous explanations found in Buddhist literature. The three characteristics which firmly anchor the practice, namely its sustained, uninterrupted and cherished character (YS I.14), are all characteristics of Buddha's practice as described by Vasubandhu. Secondly, Patañjali's division of detachment into lower and higher (YS I.15–6) is the reformulation of the Asaṅga's division of detachment in to mundane and supramundane (*laukika* and *lokottara*).

(4) Patañjali's concept of "seven-fold wisdom of the highest plane" (*saptadhā prāntabhūmiḥ prajñā*, YS II.27) could be a reflection of the notion of "seven knowledges" which an attachment-free person or the *arhat* achieves according to Vasubandhu.

(5) Scholars have faced problems in interpreting Patañjali's concept of *īśvarapraṇidhāna* (YS I.23) because they have neglected the special sense in which the term *praṇidhāna* is used in Buddhist literature. According to Mahāyāna Buddhism *praṇidhāna* stands for an ardent desire or vow or resolve. Taking a clue from this, the term *īśvarapraṇidhāna* can be interpreted as an ardent desire or resolve to become an ideal being like *īśvara*.

(6) In *Yogasūtra*, Patañjali presents at least three different formulations of the theory of meditation and meditative absorption. (1) *Samprajñāta-samādhi*—*asamprajñāta-samādhi* (YS I.17–8), (2) *samāpatti* (meditative attainment) of different kinds (with gross thoughts, with subtle thoughts, without gross or subtle thoughts)—*sabīja-samādhi*—*nirbīja-samādhi* (YS I.41–51) and (3) *dhāraṇā-dhyāna-samādhi* (YS III.1–3). The three formulations can be linked up with each other because they have Buddhist theory of meditation as their background. Accordingly, the stages in *samprajñāta-samādhi* can be correlated to the stages of *rūpadhyāna* of the Buddhist theory;[3] *asamprajñāta-samādhi* can

be correlated to *ārūpyas* of the Buddhist theory.[4] *Samādhi* and *samāpatti* are closely connected in the Buddhist theory. Similarly, they are closely connected in Patañjali's theory also. *Nirbīja-samādhi* of Patañjali can be correlated to *nirodha-samāpatti* or *saṁjñā-vedayita-nirodha* of Buddhism. Moreover, the three stages of meditation in Patañjali's theory, namely *dhāraṇā*, *dhyāna* and *samādhi*, can be correlated to the Buddhist *dhyāna* stages, namely (1) *savitarka*, (2) *nirvitarka-savicāra* and (3) *nirvitarka-nirvicāra* (with *ekāgratā* dominating), where *vitarka* (Pali-*vitakka*) is understood as "initial application of mind (or fixation of mind) to the object, and *vicāra* is understood as sustained application of mind.[5]

(7) Patañjali arranges his account of supernormal powers around the concept of *saṁyama* (integrated concentration). He defines *saṁyama* as "(Focusing) all the three (that is, *dhāraṇā*, *dhyāna* and *samādhi*) on the same object" (YS III.4). This does not make the idea very clear. Here Asaṅga's discussion (in *Śrāvakabhūmi*) of supernormal knowledges and powers can be of help, because he describes there how one cultivates and prepares one's mind for attainment of supernormal powers after attaining perfection in meditation. *Saṁyama* could be a synoptic representation of this description.[6]

(8) If we try to interpret *Yogasūtra* in the light of its Buddhist background, then sometimes we get an impression that Patañjali is trying to appropriate the Buddhist Yoga and also supersede it through this work. He seems to claim that the yogin of his Yoga system can achieve many of the high goals which a Buddhist practitioner of high order can achieve. Hence, his yogin can attain *dharmamegha* plane through Sāṅkhya-type emancipating knowledge, which *bodhisattva* of the highest plane can attain according to Mahāyāna Buddhism. Patañjali also seems to claim that the yogin of his school can attain even some of those supernormal powers which Buddhism attributes only to the Buddha. They include elephant power (YS III.24), mind-like speed (YS III.48) and omniscience (YS III.49). Whether one should take these claims seriously or not is a separate issue.

(9) While reading *Yogasūtra* one sometimes finds that the interpretation informed by the Buddhist background is more consistent or correct than the interpretation given by Vyāsa. Here are some examples: (1) Vyāsa identifies *samprajñāta-samādhi* with *sabīja-samādhi* and *asamprajñāta-samādhi* with *nirbīja-samādhi*. He interprets *bīja* as "external object" (*bahirvastu*). But if *bīja* is understood as latent impression (or *ālayavijñāna*) then *asamprajñāta-samādhi* will be *sabīja* because it is described as *saṁskāraśeṣa* (that in which latent impressions remain). Hence, *nirbīja-samādhi* is to be understood as different from and higher than *asamprajñāta-samādhi*.[7] (2) Patañjali recommends meditative cultivation of four attitudes, namely *maitrī*, *karuṇā*, *muditā* and *upekṣā* (YS I.33). In the Buddhist traditions these four attitudes are called "immeasurable" (*apramāṇa*) because they are supposed to be addressed to all beings. According to Vyāsa's interpretation, they are addressed to specific types of beings. Accordingly, *upekṣā* meant indifference, which is to be addressed

to sinful beings. But then Vyāsa has to face the problem that indifference to sinful persons cannot be "cultivated meditatively". This problem can be overcome if following Buddhist tradition; we interpret *upekṣā* as equanimity which has removal of sinfulness as its content. (3) Patañjali says that the yogin attains *dharmamegha-samādhi* when he becomes *akusīda*, even after the rise of emancipating knowledge (YS IV.29). Vyāsa has interpreted the word as "desireless and detached", which is not supported by any dictionary. However, the meaning of the term *akusīda* can be understood on the Buddhist background, which is "not lethargic" "not lacking enthusiasm", etc.

It has been argued in the preceding discussions that Patañjali in the *Yogasūtra* tries to narrate two stories of suffering and emancipation in an integral way: the metaphysical story of Sāṅkhya and the empirical–psychological story of Buddhism. The two stories have many contrasting features, which Patañjali tries to resolve by reframing Sāṅkhya categories in phenomenological terms, by making Sāṅkhya categories the objects of meditation and by giving finality and ultimacy to the Sāṅkhya metaphysics. How consistently and successfully Patañjali synthesizes the two stories remains a question. We have also seen that Vyāsa's interpretation is sometimes at discrepancy with the true import of Patañjali partly because the former does not take the Buddhist background seriously.

One of the major points this work makes is that the system called Pātañjalayoga can be understood in its true perspective only by going beyond orthodox bias and taking the Buddhist background seriously.

Notes

1 This was pointed out by Frauwallner (1984: 315–328) and followed by Larson and Bhattacharya (2008: 41).
2 As Frauwallner (1984: 317) remarks, the concept of an inner organ (*antaḥkaraṇa*) which consists of thinking (*manaḥ*), I-consciousness (*ahaṁkāra*) and the faculty of knowledge (*buddhi*) was given up in Vindhyavāsī's reconstruction of Sāṅkhya and was replaced by *manaḥ* which is supposed to perform all the functions earlier attributed to *Buddhi*.
3 See the comments on YS I.17 and also Appendix I.
4 See the comments on YS I.18 and also Appendix I.
5 See the comments on YS III.1–3 and also Appendix I.
6 For the details of Asaṅga's description, see Appendix II.
7 See the comments on YS I.18, I.46 and I.51.

APPENDIX I

Buddhist conceptions of meditation, yoga and *bodhisattva's* spiritual journey vis-à-vis Pātañjala-yoga

In this appendix, we will do three things:

(1) We will briefly survey the Buddhist theory of meditation as presented by Asaṅga and Vasubandhu and bring out its relevance for Patañjali's theory of meditation.
(2) We will take note of the term Yoga as it was used by Asaṅga in his *Śrāvakabhūmi*, which throws light on the Buddhist conception of Yoga.
(3) We will look at Asaṅga's discussion of the spiritual journey of a *bodhisattva* in which the penultimate stage (the stage immediately prior to Buddha-hood) is marked by *dharmameghā bhūmi*.

Part I: Buddhist theory of meditation

Buddhist theory of meditation, with its complexity and diversity, is at the background of Patañjali's Yoga. In Buddhism, the terms *dhyāna*, *samādhi* and *samāpatti* are used with overlapping meanings. Similarly, the concepts of *smṛti*, *dhāraṇā* and *bhāvanā* also overlap with them. Most of these terms with similar meanings are also used by Patañjali.

The major division found in Buddhist meditation theory is between *śamatha* (concentration meditation) and *vipaśyanā* (insight meditation).

There are a few aphorisms in YS which could be interpreted as referring to *vipaśyanā*.[1] But more often than not, Patañjali talks about different stages and aspects of *śamatha* meditation (or concentration meditation).

In understanding concentration meditation, Patañjali is influenced by the Buddhist theory of meditation. It may be stated here that the concentration meditation with four form meditations (*rūpadhyānas*) and four formless spheres (*ārūpyas*) is not originally Buddhist; Gotama the *bodhisattva* himself learnt it from his teachers, Āḷāra Kālāma and Uddaka Rāmaputta. But a detailed account of the theory of meditation which the *bodhisattva* received from his Brahmanical teachers is not available. We are told that the Buddha discovered a state beyond the eight stages of concentration meditation, namely cessation of perception and sensation (*saṁjñā-vedayita-nirodha* or *nirodha-samāpatti*), which is reflected in Patañjali's definition of yoga (*citta-vṛtti-nirodha*). Secondly, it looks plausible that the Buddha and his followers must have elaborated and developed the concentration meditation which they received from Brahmanical teachers. Hence it must have been the Brahmanical meditation adopted and revised by the Buddhists which must have been available to Patañjali and not the Brahmanical form of meditation itself.

In the case of Patañjali, it seems to be the theory of meditation presented by Asaṅga and Vasubandhu which was before him as a background. So, let us focus on their theory of concentration meditation.

Vasubandhu distinguishes between two conceptions of *dhyāna*: *upapatti-dhyāna* and *samāpatti-dhyāna*. *Upapatti-dhyāna* means *dhyāna* by birth-according to which different *dhyānas* are associated with birth in different worlds or planes of existence (*bhūmi*). Vasubandhu defines *samāpatti-dhyāna* as one-pointedness of wholesome mind (*kuśala-cittaikāgratā dhyānam*) (AKB VIII.1) which is attained in this world itself. Here Vasubandhu talks of eight *samāpattis*—out of which the first four are called *samāpatti-dhyāna* (which are all connected with the world of forms and hence can be called *rūpadhyāna*1). And the next four are *ārūpyasamāpattis*.

The four *dhyānas* (that is, *rūpadhyāna*1s) are stated as follows:

(1) *Dhyāna*, which possesses *vicāra* (along with *vitarka*). Here *vitarka* is understood as a gross state of mind and *vicāra* as subtle state of mind. This stage is divided into two sub-stages:

    (1a) *Vitarka+vicāra*
    (1b) *Vicāra* (without *vitarka*)

    This stage also possesses *prīti* (joy) and *sukha* (happiness).
    Out of the two sub-stages, (1b) is called *dhyānāntara* (intermediate *dhyāna*).[2]
(2) *Dhyāna*, which possesses *prīti* and *sukha* (but no *vitarka-vicāra*).
(3) *Dhyāna*, which possesses *sukha* (but no *prīti* or *vitarka-vicāra*).
(4) *Dhyāna*, without *vitarka-vicāra*, *prīti* and *sukha*.

Asaṅga has explained the process of the four *dhyānas* in a lucid way:

> What is to take recourse to *samādhi*?

(1) The fellow (i.e. the aspirant) having abandoned the five hindrances which cause subordinate and major defilements in mind, leads the life (*viharati*) by attaining the first *dhyāna*, which is excluded by sensuous desires, excluded by sinful and unwholesome characters, which possesses *vitarka*, which possesses *vicāra*, which contains joy and happiness arising from seclusion (*viveka*).

(2) This fellow (then), due to calming down of *vitarka* and *vicāra* attains the state of internal serenity (*adhyātmasamprasāda*), and due to concentration of mind leads the life by attaining the second *dhyāna*, which is without *vitarka* and *vicāra* but contains joy and happiness arising from one-pointedness of mind (*samādhi*).
(3) This fellow (then), due to detachment from joy, leads life with equanimity. He, then being mindful and vigilant, experiences that happiness by body which noble ones describe in that way. He, in this way, having become an equanimous, mindful and happy being, leads the life by attaining the third *dhyāna*.
(4) This fellow (then) abandons happiness, and because his joy and distress have ceased already, leads the life by attaining the fourth *dhyāna*, which is devoid of pleasure and pain and which possesses completely purified equanimity and mindfulness (*upekṣāsmṛtipariśuddham*). This is what it is to take recourse to *samādhi*.

(ŚB: 14–15)

Vasubandhu presents the four *dhyānas* in the same way in nutshell, which is presented in Table A1.1.

This indicates that when one passes from the first to fourth *dhyāna*, at each stage there is dropping of some factors of the previous stage and also addition of some factors. Table A1.2 makes this clear.

TABLE A1.1 Four *dhyānas* and their aspects

Dhyāna	*Number of aspects* (aṅgas)	*Aspects* (aṅgas)
First	5	*vitarka, vicāra, prīti, sukha, ekagrata*
Second	4	*prīti, sukha, adhyātmasamprasāda, samādhi*
Third	5	*sukha, upekṣā, smṛti, samprajñāna, samādhi*
Fourth	4	*asukhāduḥkha-vedanā, upekṣāpariśuddhi, smṛti-pariśuddhi, samādhi*

Sources: AK and AKB VIII.7–8

TABLE A1.2 Four *dhyānas* according to Asaṅga and Vasubandhu

Dhyāna *no.*	*Elements*	*Deletion*	*Addition*
1	*vitarka-vicāra-prīti-sukha-ekāgratā*	X	X
2	*prīti-sukha-ekāgratā-adhyātmasamprasāda*	*vitarka-vicāra*	*Adhyātmasamprasāda*
3	*sukha-ekāgratā-upekṣā-smṛti-samprajñāna*	*Prīti*	*upekṣā-smṛti-samprajñāna*
4	*ekagrata(samādhi)-asukhaduḥkha-vedanā-upekṣāpriśuddhi-smṛti-pariśuddhi*	*Sukha*	*asukhāduḥkhavedanā upekṣāparriśuddhi, smṛti-pariśuddhi*

Sources: AK and AKB VIII.7–8; ŚB: 14–15

The following points need to be noted. They are relevant for comparing the process with Patañjali's theory of meditation:

(1) In the Buddhist theory of meditation, the first *dhyāna* does not contain one-pointedness of mind (*samādhi*) as its dominant aspect; *samādhi* is included explicitly in the second *dhyāna* onwards. This explains the hierarchical order between *dhyāna* and *samādhi*.[3] That is, why joy and happiness (*prīti-sukha*) are called "caused by seclusion" (*vivekaja*) in the first *dhyāna*, whereas they are called "caused by concentration" (*samādhija*) in the second *dhyāna*.
(2) The addition of factors in the later stages of *samādhi* is reflected in Patañjali's theory also. For instance, he refers to *adhyātmaprasāda* (YS I.47), which arises in the state of *nirvicāra-samāpatti*. He also talks of *smṛti-pariśuddhi* giving rise to *nirvitarka-samāpatti* (YS I.43). Similarly, there is a close connection between *samprajñāna* occurring in the third *dhyāna* and *prajñā*.[4] The rise of *ṛtambharā prajñā* as a result of *adhyātmaprasāda* (see YS I.47–48) is parallel to the rise of *samprajñāna* in *dhyāna* 3 as a result of *adhyātmasamprasāda* in *dhyāna* 2 (see Table A1.2).
(3) Patañjali in his theory seems to divide the first *dhyāna* of the Buddhist hierarchy into two stages—*vitarka* stage and *vicāra* stage (Vasubandhu had acknowledged these stages as "first *dhyāna*", and "*dhyānāntara*" (intermediate *dhyāna*) respectively. Moreover, Patañjali seems to club the two stages of the Buddhist hierarchy together—*prīti* stage and *sukha* stage—and call it *ānanda*.

There are other aspects of Buddhist meditation which are relevant in this context.

(1) *Samāpatti* and *samādhi*: as we have seen, Buddhism distinguishes between *dhyāna-upapatti* (being born in a world/plane of existence with a *dhyāna* state) and *dhyāna-samāpatti* (attainment of *dhyāna*). The factors of *dhyāna*, *samādhi* and *samāpatti* go hand in hand. For instance the distinction between *savitarka*, *nirvitarka*, *savicāra* and *nirvicāra* applies to all the three.
(2) In the Buddhist theory, the word *vitarka* is used in various senses.

    (a) Vitarka means reasoning, thinking. Sometimes the term is used in a derogatory sense in which it refers to *akuśalavitarka* or *kliṣṭavitarka* (unwholesome thought, defiled thought).[5]
    (b) But *vitarka* does not always mean "immoral thinking". Sometimes it means "gross thinking" as against *vicāra* which means subtle thinking, "*cittaudārikatā vitarkaḥ cittasūkṣmatā vicāraḥ*" (AKB II.33a). Patañjali, too, suggests this (YS I.44). Vyāsa explains *vitarka* and *vicāra* in the same way.[6] The difference between *vitarka* (gross thought) and *vicāra* (subtle thought) seems to be that *vitarka* contains passions—whether moral or immoral—in a heterogenous and disturbing way.[7] *Vicāra*, on the other hand, is subtle thought in the sense that it is homogenous and non-disturbing.

(c) In Pali tradition, *vitarka* (Pali: *viṭakka*) and *vicāra* have a different significance in the context of the theory of meditation. T. W. Rhys Davids and William Stede, in Pali English Dictionary, give one of the meanings of *vitakka* as "initial application" and *vicāra* as "sustained application". This meaning, according to them, gives *vitakka* the characteristic of fixity and steadiness and *vicāra* that of movement and display: "*vitakka* is the directing of concomitant properties towards the object; *vicāra* is the continued exercise of the mind on that object". This explanation brings the *vitarka* stage of the first *dhyāna* close to what Patañjali calls *dhāraṇā* (fixed attention), as both of them are about fixing or steadiness of mind. Similarly, the *vicāra* stage of the first *dhyāna* (which Vasubandhu terms as *dhyāntara*) is close to what Patañjali calls *dhyāna*, as both of them are about free and continuous movement of mind (or sustained application of mind) with respect to the object. From the second *dhyāna* onwards, we have in Buddhism what Patañjali calls *samādhi*—something like absorption of mind in the object. That is why Patañjali's description of "*nirvitarkā samāpatti*" (YS I.43), namely "*svarūpaśūnyā iva arthamātra-nirbhāsā*" (that which is devoid of its own nature and in which only the object appears), is repeated in his definition of *samādhi* (YS III.3). In fact, identification of "*nirvicārā samāpatti*" with "*samādhi*" would have been more appropriate.

Ārūpyas

The four *dhyānas* stated earlier are called *rūpadhyāna*. The objects of these meditative practices have a tangible form. In this sense, they belong to the world of forms (*rūpadhātu*). Having practiced the *rūpadhyāna* sufficiently, the practitioner may develop detachment towards the formed objects (the objects of the *rūpa* world) and concentrate on formless objects. By this the practitioner reaches a higher stage called *arūpa* (formless) or *ārūpya* (formlessness). In formless stage, one does not concentrate on an object. In fact *ārūpyas* cannot be called states of concentration in the ordinary sense of the word. They are formless states of consciousness or spheres of consciousness (*āyatanas*) with emphasis on different aspects of it. There are four such *ārūpyas*:

(1) Sphere of infinite space (*ākāśānantyāyatana*)
(2) Sphere of infinite consciousness (*vijñānānantyāyana*)
(3) Sphere of nothingness (*ākiñcanyāyatana*)
(4) Sphere of neither perception nor non-perception (*naiva-saṁjñā-nāsaṁjñāyatana*)

A little analysis can reveal that these different spheres are different aspects of object-less consciousness. "Space" in Buddhism is not a positive substance but just vacuum. Hence, the sphere of infinite space is not consciousness on any positive thing. The second sphere is infinite consciousness, which is not an "object" of consciousness but consciousness itself. The third sphere is the sphere of nothingness, which is obviously not a "positive object" of consciousness. Vasubandhu suggests that these three are not objects (*ālambana*) of meditative consciousness, but they result from the mental applications of that kind.[8] The fourth sphere is rather tricky. It is called neither perception nor non-perception. Vasubandhu says that it is so-called because it is too weak to be called perception so that it is as good as non-perception.[9] Perhaps it can be described as perception which is not perception of anything, that is, an objectless perception. Perhaps what Patañjali calls "other type of *samādhi*" (what is popularly termed as *asmprajñāta-samādhi*) in YS I.18 is equivalent to what the Buddhists regard as a formless meditative trance. Patañjali describes it as *saṁskāraśeṣa* (that is, that in which latent impressions are remaining). This implies that though *asamprajñāta-samādhi* is supposed to be higher than the *samprajñāta-samādhi*, it is still not perfect or ultimate.

The *ārūpya* (formless) stage of meditative trance shares this feature, too, with *asamprajñāta-samādhi*. One indicator of this deficiency in formless meditation is the presence of fetters (*saṁyojanas*). Fetters are inflows or impurities which keep one bound to the worldly existence. Buddhists accept in all ten fetters:

(1) Dogma of the self (*satkāyadṛṣṭi*)
(2) Attachment to mere rules and rituals (*śīlavrataparāmarśa*)
(3) Doubt (*vicikitsā*)
(4) Excitement for serious pleasure (*kāmacchanda*)
(5) Hatred (*vyāpāda*)

(The five fetters are called lower fetters as they belong to the world of sensuality (*kāmadhātu*). The other five fetters are upper fetters which operate in the higher worlds—namely the worlds of form (*rūpa*) and formlessness (*ārūpya*)). The latter five fetters are as follows:

(6) Lust for the world of forms (*rūparāga*)
(7) Lust for the formless world (*arūparāga*)
(8) Excitement (*auddhatya*)
(9) Conceit (*māna*)
(10) Misconception (*avidyā*)[10]

In *ārūpya*-stage, *rūparāga* (lust for *rūpa* world) does not remain, but lust for *arūpa*-world and other higher fetters may remain.

When one develops detachment even to the formless objects, one can reach the highest meditative trance. In Buddhism this stage is termed as *saṁjñā-vedayita-nirodha* (cessation of perception and feeling) or *nirodha-samāpatti* (meditative attainment of cessation). In Patañjali's scheme this stage would be beyond *asamprajñāta-samādhi*. Patañjali calls it *nirbīja-samādhi* (seedless trance).

Part II: Asaṅga's conception of yoga

The word *yoga*, which is a noun derived from the verb *yuj* (which means, "to join"), means association. Whether an association is good or bad depends on that with which one is associated. Hence, the word *yoga* can have a positive or negative connotation depending upon the context. In the Pali Buddhist literature as well as Sanskrit Abhidharma literature the word *yoga* is used as a technical term meaning unwholesome association.[11] Asaṅga (ADS, 47) and Vasubandhu (AKB VIII.4) refer to four types of unwholesome associations (four types of *yoga*), namely *kāmayoga*, *bhavayoga*, *dṛṣṭiyoga* and *avidyāyoga*. Asaṅga, after giving the list of the four unwholesome associations, says that they are called *yoga* because they are opposed to dissociation (*visaṁyoga*) and they are contrary to purification.[12] On the other hand, in *Upaniṣads* and Brahmanical literature, the words *yoga* and *yogin* were used more consistently in a positive sense, as those of "association with spiritual goal" and "spiritual practitioner" respectively. Though in the early Buddhist literature also such positive usages are found, they are occasional and exceptional. It can be argued that the term *yoga* in its positive sense was given central status in the Buddhist soteriology for the first time with the rise of the Yogācāra school of Buddhism. With the development of the Yogācāra school of Buddhism, the term *yogācāra* was associated with idealism or "mind-only thesis". But initially the word *yogācāra* did not have this connotation. The word "*yoga*" in the compound word "*yogācāra*" had neither a negative connotation of unwholesome association, nor did it have the idealistic connotation. It had the connotation of "spiritual practice" or "association with spiritual goal", as it had in the Brahmanical tradition.

Asaṅga in *Śrāvakabhūmi* explains the concept of Yoga by throwing light on various aspects of it. He asks the following questions and answers them. Here is a summary:

Questions

(1) What is a fall from *yoga* (*yogabhraṁśa*)?
(2) What is *yoga*?
(3) What are the functions of *yoga* (*yoga-karaṇīya*)?
(4) How many types of *yoga*-practitioners (*yogācāra*) are there?
(5) What is cultivation of *yoga* (*yogabhāvanā*)?
(6) What is the fruit of the cultivation of *yoga* (*bhāvanāphala*)?

Asaṅga's answers were briefly as follows:

(1) **What is a fall from *yoga* (*yogabhraṁśa*)?**

A fall from *yoga* is of four kinds:

(a) Absolute (*ātyantika*) fall from *yoga* belongs to the beings, incapable of *yoga*.

(b) Temporary (*tāvatkālika*) fall from yoga belongs to those having yogic eligibility, who are bound to attain emancipation, but due to weakness of conditions are driven away from *yoga* and are emancipated by pursuing it again.
(c) Fall from *yoga* after attainment (*prāptiparihāṇika*) belongs to the beings who attain *yoga* but lose the achievements (in terms of knowledge, vision and conduct).
(d) Fall from *yoga* caused by wrong belief (*mithyāpratipattikṛta*) belongs to the beings who are intelligent but follow a wrong path because of wrong instructions from wrong teachers.

(2) What is *yoga*?

Answer: *Yoga* is four-fold (*caturvidha*) (that is, the practice of *yoga* has four aspects): faith (*śraddhā*), desire (*chhanda*), energy (*vīrya*) and means (*upāya*).

(a) *Śraddhā* is of the form of trust or faith or confidence (*prasāda*) which is based on either rational thinking on *dharma* or on resolved faith in great beings.
(b) *Chhanda* (desire) is of four kinds: (i) desire to achieve liberation; (ii) desire to know through inquiry; (iii) desire to develop good conduct, self-control, consciousnesses etc. and (iv) desire to maintain continuity (*sātatyaprayogatā*), to practice with devotion (*satkṛtya-prayogatā*).
(c) *Vīrya* (energy) is of four kinds: (i) energy to learn (by hearing), (ii) to think over it, (iii) to meditate and (iv) to clear (remove) the veils.
(d) *Upāya* (means) is also of four kinds: (i) presence of mindfulness (*sūpasthitasmṛtitā*) based on control over conduct and control over senses, (ii) having achieved mindfulness, protecting mind from heedlessness (*apramāda*), (iii) practicing wholesome actions (*kuśalānāṁ dharmāṇāṁ niṣevaṇā*) for concentration of mind in an inner object (*adhyātmaṁ śamathayogaḥ*) and (iv) development of insight (*vipaśyanā*).

(3) What is to be done through *yoga* (*yoga-karaṇīyam*)?

Answer: Four things are to be done with the help of *yoga*: (i) stoppage of the (*kārmic*) basis (*āśrayaṇirodha*), (ii) transformation of the (kārmic) basis (*āśraya-parivarta*), (iii) knowing the object fully (*ālambanaparijñāna*), (iv) engaging with the object (*ālambanābhirati*).

(4) How many types of *yoga*-practitioners are there?

Answer: There are three types of *yoga*-practitioners (i) beginner (*ādikarmika*), (ii) experienced practitioner (*kṛtaparicaya*[13]) and (iii) master (who has transcended the stage of mental application) (*atikrāntamanaskāra*).

(5) What is the cultivation of *yoga* (*yogabhāvanā*)?

Answer: It is of two kinds: (a) cultivation of perception (*saṁjñābhāvanā*) and (b) cultivation of enlightenment-promoting factors (*bodhipakṣyā bhāvanā*).

(a) **Cultivation of perception** includes perceiving faults in the objects belonging to lower realms, perceiving abandonment, detachment and cessations in the things eligible for abandonment etc., perceiving objects of concentration meditation (*śamathapakṣyā saṁjñā*) and perceiving objects of insight meditation (*vipaśyanāpakṣyā saṁjñā*).

(b) **Cultivation of enlightenment-promoting factors** means the practice of thirty-seven pro-enlightenment factors (*bodhipakṣya-dharma*). They include four mindfulnesses, four right abandonments, four supernormal powers, five faculties, five powers, seven enlightenment factors (*bodhyaṅga*) and eight limbs of the noble path.

(6) What is the fruit of the cultivation of (*yoga*)?

Four fruits of practicing monk's life (*śrāmaṇya*) are the fruits of cultivation of *yoga*. They are

(i) Fruit of stream-entrance (*srota-āpatti-phala*)
(ii) The fruit of once-returner (*sakṛdāgāmiphala*)
(iii) The fruit of non-returner (*anāgāmiphala*)
(iv) The fruit of the accomplished one (*arhatphala*)

The conception of *yoga* which emerges from this description is that of the practice of self-discipline towards the moral-spiritual goal which is to be achieved within the framework of Śrāvakayāna. Asaṅga here is adopting the general positive concept of *yoga* and applying it in the framework of Śrāvakayāna Buddhism.

The description of the *yoga*-related terms (including the term "*yogācāra*") also suggests that the early Yogācāra was non-idealistic.[14]

Asaṅga's conceptions of *yoga* and *yogācāra* in this way must have introduced an alternative to the traditional Brahmanical *yoga* system which was not sufficiently systematic and elaborate. Asaṅga's formulation of Yogācāra as a full-fledged system of *yoga* (which was based on the Buddhist's doctrines of non-eternality and no-self) must have posed a challenge before Patañjali who strongly believed in Sāṅkhya metaphysics (which accepted eternality and the *puruṣa*). This might have led him to construct a system of *yoga* by incorporating the Buddhist empirical content in the Sāṅkhya metaphysical framework.

Part III: Asaṅga on *bodhisattva's* spiritual development vis-à-vis *dharmameghā bhūmi*

Buddhism is often divided into three paths (*yāna*), namely Śrāvakayāna (the path of disciplines), Pratyekabuddhayāna (the path of individually awakened ones) and Bodhisattvayāna (the path of beings destined to be awakened). Though Patañjali while incorporating the elements of the Buddhist theory of meditation in his reconstruction of Yoga mainly uses the terms, concepts and doctrines of Śrāvakayāna, sometimes he uses the terms and concepts used in Bodhisattvayāna as well. For instance, in the last chapter of *Yogasūtra*, Patañjali talks about the

TABLE A1.3 Correlation between thirteen *vihāras* and ten *bhūmis*

Vihāra *no.*	*Name of the* vihāra	*Name of the* bhūmi *reached in the respective* vihāra
1	*Gotravihāraḥ*	—
2	*Adhimukticaryāvihāraḥ*	—
3	*Pramuditavihāraḥ*	*Pramuditā*
4	*Adhiśīlavihāraḥ*	*Vimalā*
5	*Adhicittavihāraḥ*	*Prabhākarī*
6	*Bodhipakṣyapratisaṁyuktaḥ adhiprajñavihāraḥ*	*Arciṣmatī*
7	*Satyapratisaṁyuktaḥ adhiprajñavihāraḥ*	*Sudurjayā*
8	*Pratītyasamutpādapratisaṁyuktaḥ adhiprajñavihāraḥ*	*Abhimukhī*
9	*Sābhisaṁskāraḥ sābhogaḥ nirnimitto vihāraḥ*	*Dūraṅgamā*
10	*Anābhogaḥ nirnimitto vihāraḥ*	*Acalā*
11	*Pratisaṁvidvihāraḥ*	*Sādhumatī*
12	*Bodhisattvānāṁ paramo vihāraḥ*	*Dharmameghā*
13	*Tāthāgato vihāraḥ*	*Pratiṣṭhā*

Source: BSB: 217–44, 281

Note: The names of ten *bhūmis* are due to *Daśabhūmīśvara-sūtra*. But the name of the last *bhūmi*, *pratiṣṭhā*, might have been given by Asaṅga himself.

highest meditative trance the yogin attains, namely "*dharmamegha-samādhi*", which reflects the highest plane of the *bodhisattva*'s spiritual journey according to Bodhisattvayāna. Here Patañjali might have Asaṅga's *Bodhisattvabhūmi*, which is supposed to be the fifteenth section of his voluminous work *Yogācārabhūmi*, as his background.

In the chapter called "Vihāra-paṭala" of *Bodhisattvabhūmi*, Asaṅga discusses the spiritual journey of a *bodhisattva* from his membership in a *bodhisattva* family (*gotra-vihāra*) to leading the life of the Buddha (*tāthāgata-vihāra*). The journey consists of thirteen stages which Asaṅga calls *vihāras* ("ways of leading life", in this case, "stages in spiritual life"). Out of them, Asaṅga correlates the ten stages with ten *bhūmis* as they are explained in Mahāyāna canonical work *Daśabhūmisūtra*. Asaṅga devotes the last chapter of *Bodhisattvabhūmi*, namely "Pratiṣṭhā-paṭala", for discussion of the final stage of this journey, namely Buddha-hood.

The list of thirteen *vihāras* and their correlation with ten *bhūmis* is seen in Table A1.3.

A brief description of the spiritual stages

(1) *Gotra-vihāra* (natural eligibility to become a member of the *bodhisattva* family): *Bodhisattva* in this stage leads virtuous life by temperament/nature, not by force. He is undisposed to favour impurities or to destroy wholesome roots. He contains the potential of a *bodhisattva*.

(2) *Adhimukticaryāvihāra* (inclination to lead the life of *bodhisattva*): *Bodhisattva* in this stage begins to lead the life of a *bodhisattva* consciously, but half-heartedly, with many doubts, impurities and imperfections.

(3) *Pramuditavihāra* (joyfully joining the *bodhisattva* career): *Bodhisattva* in this stage takes *bodhisattva* vow and becomes "born" in *bodhisattva* family. Having perceived the spiritual achievements of other *bodhisattvas*, he experiences joy and happiness. Makes different types of resolves (*bodhisattva-praṇidhāna*) to perform *bodhisattva* activities and achieve spiritual goals. (Asaṅga here mentions ten different resolves.) *Bodhisattva* in this stage attains "*pramuditā-bhūmi*".

(4) *Adhiśīlavihāra* (engaging in moral conduct): *Bodhisattva* in this stage removes all the roots of immorality (*akuśalamūla*) and engages in the course of moral conduct. He also orients other beings to do so. *Bodhisattva* in this stage attains "*vimalā bhūmi*". It is called so because the dirt of immorality is removed from it.

(5) *Adhicittavihāra* (engaging in meditational practices): *Bodhisattva* in this stage focuses on mental purification with the help of meditations, meditative absorptions and meditative attainments (*dhyāna-samādhi-samāpatti*). *Bodhisattva* in this stage attains *prabhākarī bhūmi*. It is called so because it is influenced by the light (*āloka, prabhā*) of the purity of mind generated by meditative concentration.

(6) *Bodhipakṣyapratisaṁyuktaḥ adhiprajñavihāraḥ* (engaging in wisdom regarding pro-enlightenment factors): This stage and the next two stages are engaged in wisdom (*prajñā*). In this stage it is the wisdom concerning pro-enlightenment factors. There are thirty-seven pro-enlightenment factors.[15] *Bodhisattva* in this stage practices them as they are taught by the Buddha. In this stage he attains the plane called *arciṣmatī bhūmi*. It is called so because in this plane the pro-enlightenment factors become the "flames of knowledge" (*jñānārcirbhūtāḥ*).

(7) *Satyapratisaṁyuktaḥ adhiprajñavihāraḥ* (engaging in wisdom regarding the four noble truths): In this stage the *bodhisattva* develops understanding about all the aspects of the four noble truths, namely suffering, its cause, cessation and the path leading to the cessation. In this stage he also masters different skills and arts useful for helping the living beings. *Bodhisattva* in this stage attains the *bhūmi* called *sudurjayā* (undefeatable). It is so called because the certain knowledge of the truths cannot be defeated.

(8) *Pratītyasamutpādapratisaṁyuktaḥ adhiprajñavihāraḥ* (engaging in wisdom regarding dependent origination): In this stage the *bodhisattva* realizes how the worlds arise and cease dependently (that is, relative to conditions) and hence the threefold door of liberation becomes available to him: the empty, the uncaused, the aimless (*śūnyam animittam apraṇihitam*). *Bodhisattva* in this stage attains the *bhūmi* called *abhimukhī*. It is so called because it is favourable to unattached knowledge (*asaṅga-jñāna*) and to perfection of wisdom (*prajñāpāramitā*).

(9) *Sābhisaṁskāraḥ sābhogaḥ nirnimitto vihāraḥ* (engaging in the unconditioned with efforts): This is the stage of transition from some element of impurity to extreme purity, the stage in which the perfections and the pro-enlightenment factors attained in previous stages attain completion, all the defilements are eliminated and one is ready to enter the next stage which stands for extreme purity (*ekāntaviśuddha*). In this stage the *bodhisattva* attains the *bhūmi* called *dūraṅgamā* (far-reaching) because he attains perfection in *bodhisattva* conduct in this stage.

(10) *Anābhogaḥ nirnimitto vihāraḥ* (engaging in the unconditioned without efforts): In this stage the *bodhisattva* attains the purest and highest form of faith (*kṣānti*). He attains various masteries (vaśitā) and performs supernormal actions easily. The defects, the impressions of actions and mental efforts present in the previous stage, cease in this stage. Because of that his mind becomes firmly established on the path. Hence the *bhūmi* he attains in this stage is called *acalā* (immovable).

(11) *Pratisaṁvidvihāraḥ* (developing analytical knowledge of *dharma*): In this stage the *bodhisattva* develops the analytical knowledge of *dharma* and the capacity to preach it. In this stage the *bodhisattva* attains the *bhūmi* called *sādhumatī*. The *bhūmi* is so called because *bodhisattva*'s mind in it is filled with the thought of the well-being and happiness of all beings.

(12) *Bodhisattvānāṁ paramo vihāraḥ* (the highest stage of a *bodhisattva*): In this stage *bodhisattva* attains all the wealth of enlightenment (*bodhisambhāra*) and becomes eligible to become the Buddha (*dharmarājatvārha*). In this stage there is abandonment of all the veils of defilements, latent impressions and dormant defilements.[16] In this stage the *bodhisattva* attains the highest *bhūmi* called *dharmameghā*. The word *dharmameghā* is the adjective of *bhūmi*. So it can be deciphered as *Bahuvrīhi* compound and can be taken to mean "that in which virtue (or goodness, *dharma*) showers like a cloud".

(13) *Tāthāgato vihāraḥ* (the stage of the Buddha): In this stage *bodhisattva* has completed his journey as *bodhisattva* and has become Buddha (Tathāgata). His *kleśāvaraṇa* was already destroyed in the last stage. In this stage his *jñeyāvaraṇa* also gets destroyed. Hence, he becomes omniscient. He has attained all the ten powers (*daśa-bala*) of the Buddha and all the unique properties (*āveṇika-buddha-dharma*). Asaṅga calls this stage *tāthāgatī bhūmi*, and also, *pratiṣṭhā* (established state).[17]

Notes

1 Patañjali refers to *ṛtambharā prajñā* (YS I.48, 49) which is analogous in content with *vipaśyanā*. Similarly, Tandon (1998: 36) has interpreted YS IV.25 as referring to *vipassanā*- "*viśeṣadarśina ātmabhāvabhāvanāvinivṛttiḥ*" ("The notion of Self ceases for him who sees in a special way"). Tandon remarks that this seems to be an unmistakable reference to *vipassanā*.

2 See AKB I.32c, YAB: 74.

3 For Vasubandhu's distinction between *dhyāna* and *samādhi*, also see AKB VIII.1d. Accordingly *dhyāna* is a more general concept, whereas *samādhi* is a special kind of *dhyāna*.

4 Vasubandhu in AK VIII.8 uses the word *prajñā* as a synonym of *samprajñāna*.

5 As Asaṅga introduces the notion of *vitarka* in the context of "*vitarkabādhana*": "*tatra vitarkaḥ kāmavitarkādayaḥ kliṣṭā vitarkāḥ*" (ŚB: 399).

6 "*vitarkas cittasyālambane sthūla ābhogaḥ, sūkṣmo vicāraḥ*"—YB I.17 (*vitarka* is a gross engagement of mind with the object, the subtle one is *vicāra*).

7 As Asaṅga refers to eight *vitarka*s such as *kāmavitarka* and describes them as "*aunmukta-saṅksobhakarāḥ cittasya*" (*vitarka*s as causing unregulated-ness and disturbances of mind) (ŚB: 415).

8 "Do these formless spheres such as infinite space have the objects such as space? No. How then? . . . One gets engaged in these formless spheres after making mental applications of that kind, namely, 'The space is infinite', 'Consciousness is infinite', and 'There is nothing' respectively. That is why they are called as that" (AKB VIII.4abc).

9 AKB VIII.4cd.

10 Source: AK and AKB V.43–45.

11 In early Jaina soteriology the word "*yoga*" was also used in the sense of "operation of the body speech and mind", which is the cause of the inflow of karma (which the soul has ultimately to get rid of). See TAS 6.1–2. The later Jaina thinkers adopted the Brahmanical concept of *yoga* meaning the practice of spiritual self-disciplining. Hence the authors like Haribhadrasūri (eighth century CE) wrote *Yogabindu* and Hemacandra (twelfth century CE) wrote *Yogaśāstra* on similar lines with Patañjali.

12 "*visaṁyogaparipanthakaro yogārthaḥ viśuddhiviparyayataḥ*" (ADS: 47).

13 Vasubandhu uses the term "*kṛtaparijaya*", which can be interpreted as one who has mastered the method (AKB VI.9d).

14 This is in tune with Schmithausen's view that "Like almost the whole of the *Yogācārabhūmi* and even many parts of other early Yogācāra texts, the *Initial Passage* does not show any trace of idealism or spiritualism, but on the contrary plainly contradicts such a position" (Schmithausen, 1987: 32).

15 Thirty-seven *bodhipakṣya-dharma*s consist of four *smṛtyupasthānas*, four *samyak-pradhānas*, four *ṛddhi-pāda*s, five *indriya*s, five *balas*, seven *bodhyaṅgas* and eight limbs of the *aṣṭāṅgika-mārga*. (Sanskritization from the Pali version as in Davids and Stede, Pali English Dictionary, the entry on *Bojjhaṅga*.)

16 "*parame punar vihāre sarvakleśasavāsanānuśayāvaraṇaprahāṇaṁ veditavyam*" (BSB: 243).

17 "*ayaṁ sa tāthāgato vihāras tāthāgatī bhūmiḥ pratiṣṭhetyucyate*" (BSB: 281).

APPENDIX II

Asaṅga on forms of supernormal knowledge and powers (*abhijñā*s and *ṛddhi*s)

The notion of *saṁyama* ("integrated concentration") is central to Patañjali's theory of supernormal powers. He explains *saṁyama* as triply focused concentration which contains fixed attention, meditation and meditative absorption (*dhāraṇā*, *dhyāna* and *samādhi*) on a single object. It can be called the meditational method of attaining a supernatural power.[1]

Patañjali's approach seems to be partly based on the Buddhist approach reflected in the Buddhist works such as *Śrāvakabhūmi* of Asaṅga. Asaṅga in *Śrāvakabhūmi* (pp. 460–68) gives a detailed description of different supernormal powers and explanation of the meditational technique which leads to them. Here I am giving a translation of the relevant section which can be used as a point of reference for understanding Patañjali's account in a clearer way.

Translation

The five forms of supernormal knowledge (*abhijñā*) are accomplished by taking recourse to meditation (*dhyāna*). How are they accomplished? The meditator, practicing it, achieves (first) an extremely pure meditation (*pariśuddha-dhyāna*). Then on the basis of the pure meditation, he applies his mind (*manasi kurvan*) to the object falling in the domain of supernormal knowledge, which he hears, picks up (*udgṛhīta*) and appropriates (*paryavāpta*)—whether it be the object of supernormal achievement (*ṛddhiviṣaya*) or recollection of prior births (*pūrvenivāsa*), divine ear, knowledge of deaths and rebirths (of other beings) or knowledge of (others') mental states. He applies the mind by the single-pointed mental application (*samāhitabhūmikena manaskāreṇa*) and through that he accurately knows the meanings and the phenomena (*arthapratisaṁvedī*, *dharmapratisaṁvedī*). When he modifies his mind (*abhisaṁskurvato*) many times, due to this multiple activity, the time arises; the condition arises when the five forms of supernormal knowledge resulting from meditative practice take place.

He, while knowing accurately the meanings and phenomena, meditates on twelve ideations (*saṁjñā*)[2] in order to accomplish all the forms of supernormal knowledge. The twelve ideations are (1) lightness (*laghu*), (2) softness (*mṛdu*), (3) space-element (*ākāśadhātu*), (4) the fusion of body and mind (*kāyacittasamavadhāna*), (5) conviction/resolution (*adhimukti*), (6) the reflection on prior actions occurring in sequence (*pūrvānubhūtacaryānukramānusmṛti*), (7) the notion of different assembled sounds of words (*nānāprakāraśabdasannipātanirghoṣa*), (8) the sign of appearance (*avabhāsarūpanimitta*), (9) transformation of appearance caused by defilements (*kleśakṛtarūpavikāra*), (10) deliverance (*vimokṣa*), (11) mastery (*abhibhvāyatana*) and (12) a holistic object (*kṛtsnāyatana*).[3]

[(1) Lightness]

In the ideation of lightness, one resolves oneself to be light-weighed such as a piece of cotton (*tūla* or *karpāsa*) or one resolves oneself to be in the volume of air and directs oneself here by a conviction-based mental application (*ādhimokṣikeṇāpi manaskāreṇa*). For example, from a bed (to seat and) from seat to bed. Similarly, from bed (to a grass-bed and) from grass-bed to bed.

[(2) Softness]

Among them, consider the ideation of softness. One affirms oneself to be soft-bodied. For example, like a silk or a bunch of hair or a piece of cloth.[4] This is the ideation of softness which nourishes and supports the ideation of lightness. The ideation of lightness, as supported (by the ideation of softness) becomes greatly increased and abundant.

[(3) Space-element]

The ideation of space-element is that by which one affirms one's own lightness and softness. If such a person wants to go anywhere, then all the interim obstacles to the act of going belonging to a material thing, are resolved as "space", by the conviction-based mental application.

[(4) Fusion of mind and body]

The ideation of the fusion of mind and body is that by which one establishes mind in the body or body in the mind by which his body becomes lighter, softer, more active and more glittering and it becomes connected with the mind, invariably related with the mind and regulated by the mind.

[(5) Conviction/resolution]

The ideation of conviction/resolution is that by which a distant thing is resolved as proximate, proximate thing is resolved as remote, a minute thing is resolved as gross, a gross thing as minute, earth as water, water as earth, in this way each gross element is resolved as another mutually. By extension, things are created by resolution, a material creation or a verbal creation.

Having accomplished the five ideations by meditating on them in this way, one experiences different kinds of supernormal powers; as follows:

(1) While being one, one shows oneself to be many; one does that by the ideation of "creation through resolution" (*ādhimokṣikayā nairmāṇikyā saṁjñayā*).
(2) After showing oneself as many one again becomes one; one does this by the ideation of "resolution of disappearance of a creation" (*nirmāṇāntardhāyikayā adhimuktisaṁjñayā*).
(3) He goes through a wall, through a mansion by the body which does not stick to them.
(4) He sinks into the ground and comes out of it as if through water.
(5) He walks on water without dividing the stream, as if on the ground.
(6) With crossed legs he traverses the sky like a winged bird.
(7) Or he touches by his hand the sun and the moon and strokes them as if they are great persons with great supernormal power.
(8) He brings under control (the higher worlds) up to Brahma world by his body. He does all this by his conviction-based cognition possessed with the notions of lightness, softness, space-element and the fusion of mind and body. They are to be understood as applicable suitably.
(9) Among them the act of bringing Brahma world under control is of two kinds: either he brings it under control by going there or by resolving to be there.
(10) With reference to the worlds lower than Brahma world the beings have a desire to wander there and assume the form of one of the derived matters (*upādāyarūpa*).

[(6) Reflection on the sequence of the prior conducts]

Among them the ideation of the reflection on the sequence of the conducts which are experienced before, is such that by it wherever his memory goes, starting from childhood, is not obstructed. One mentally perceives (*sañjānāti*) through reflection, the whole of one's experienced prior behavior elaborately (starting from childhood, for example)—where one went, stayed, sat, slept, in a gross way in a sequence without taking leaps. By associating it with meditation, as a result of meditation, he remembers his varied prior abodes of existence, as per the transmigration, elaborately with all the details.

[(7) The notion of uproar caused by the collocation of different types of sounds (divine ear)]

In a certain village or a township or a market place or assembly or a gathering or a spacious large house or an inner apartment (*Avavaraka*) a mixed and diverse uproar of an assembled mob of people who sit together or gather together, is produced, which is called confused sound (*Kalakalaśābda*) (that can be the object of meditation). Or when a large flowing river produces uproar, taking it up as the object one applies one's mind to it on a concentrated plane, and earnestly applies his mind (*ābhoga*)[5] to the sounds noble or ignoble, heavenly or human, close or remote. One achieves divine ear as the result of meditation by multiple practice of it. By that he hears divine or human sounds whether they are remote or close.

[(8) The ideation of appearance of visible from as the object (divine eye)]

Like in the earlier case,[6] one takes up a visible object and applies mind to it. This is the ideation of appearance of visible form as an object. By applying meditation to it one achieves the knowledge of deaths and births of beings as a result of that meditation. With the help of this pure divine eye, by continuing it up to one's death, when one goes to heaven, one is reborn in heaven among the gods.

[(9) The ideation of changes in the physical appearance caused by defilements (the knowledge of other's mental states)]

By this (technique), one closely observes and distinctly identifies the physical states of the beings who have attachment, hatred or delusion, and whose minds are covered by the (primary) defilements and secondary defilements such as burning, shamelessness, absence of moral fear coupled with anger and malice. If a person has attachment, then physical states and physical changes occur in him like this—his organs get excited and smile appears on face. The physical state and physical change in a hateful person occur like this—colourless face and convulsive (*gadgada*) voice; also rise of eyebrows. Similarly, the physical state and physical change in a deluded person occur like this—dumbness, inactiveness (*apratipadyanatā*)[7] towards comprehension of meaning and a tendency to make natural or non-natural speech acts.

(Having observed) whatever the physical state and physical change occur with these and such types of expressions, in the person possessed with (different primary and secondary defilements) up to immodesty and shamelessness, he takes it up as object of mental application. By meditating on it for many times, one experiences the result of meditation, namely the knowledge of the mental states (of others). By that one knows the gross thinking and subtle thinking minds of other beings and other persons as they occur.

[(10) Deliverances (11) the sense of overcoming (12) holistic objects]

These ideations (as the means of attaining supernormal powers) are to be understood in the way explained earlier. That is, "in the concentrated stage of mind" etc.[8] By such a meditational practice, one accomplishes noble supernormal power—which transforms the thing, creates a thing and resolves (the existence of one thing in place of another thing). This (supernormal power also) involves absence of strife, knowledge resulting from resolve and the four types of accurate knowledge (*pratisaṁvid*). They are as follows: accurate knowledge of phenomena, accurate knowledge of meaning, accurate knowledge of explanations and accurate knowledge of presence of mind.[9]

[Asaṅga then distinguishes between worthy and unworthy (or noble and ignoble: *ārya* and *anārya*) supernormal powers.] The noble supernormal power has these distinctive features: When one transforms something or creates an object or whatever one resolves to be by noble supernormal power, the thing becomes (genuinely) that (thing) and not otherwise. The required function can be performed by all that. But by ignoble supernormal power, a thing does not become that (genuinely) but creates only an appearance of its false form. In this way the accomplishment of five supernormal cognitions and extraordinary noble qualities by multiple mental applications on these twelve notions, is to be understood.

*

Observations

Asaṅga explains in this passage the special meditational method of achieving supernormal cognitions and powers which consists of the following factors:

(1) Extremely pure meditation (*pariśuddha-dhyāna*)
(2) Concentrated application of mind (*samāhitabhūmikena manaskāreṇa*; *samāhita* means which contains *samādhi*)
(3) Accurate knowledge of meanings and phenomena (*arthapratisaṁvedī*, *dharmapratisaṁvedī*)
(4) Application of the mind to objects many times (*bahulīkārānvaya*)
(5) Resolve (*adhimokṣa*)

Patañjali, while explaining such a method in *Yogasūtra*, talks about integrated concentration (*saṁyama*) which he defines as practicing all the three (*dhāraṇā*, *dhyāna* and *samādhi*) on a single object. This appears to be an abridged and simplified version of the detailed technique explained by Asaṅga. Secondly, Asaṅga's theory attaches special importance to the concept of resolve or conviction (*adhimokṣa*). Resolve seems to operate as a general condition of all supernormal cognitions/powers and specific types of resolves are supposed to be instrumental to specific forms of supernormal knowledge/powers. But the notion of resolve seems to be completely absent from Patañjali's theorization.

Of course, Asaṅga was not the founder of the theory of *abhijñā*s and *ṛddhi*s. He seems to have derived his theory from the Pali sources such as *Dīghanikāya* (*Pāṭikavaggo*) and *Saṁyuttanikāya* (*Ayogulasuttam*), but it is highly probable that he elaborated and systematized the original theory. Patañjali on the other hand seems to appropriate most of the Buddhist claims and adds many more, but does not go into the details of the method of achieving those *abhijñās* and *ṛddhis*. He refers to integrated concentration (that is, triple concentration) as a key to the supernormal knowledge or power related to any object. He uses the notion of *saṁyama* as if it is a key to open magically all the gates to supernormal powers and cognitions.

In the Buddhist theory, *ṛddhi* is included under *abhijñā*. In fact, *abhijñā* and *ṛddhi* refer to two different kinds of powers. *Abhijñā* refers to supernormal cognitive power (*abhi+jñā*) whereas *ṛddhi* (which literally means excess, achieving something in excess) refers to supernormal power to do something or to become something. But the theory clubs the two kinds of powers together under the common term *abhijñā*.

To sum up, the supernormal powers mentioned by Asaṅga are as follows:

Cognitive powers: (*abhiñās* other than *ṛddhis*)

(1) Knowledge of one's own prior births/abodes (*pūrvanivāsānusmṛti*)
(2) Knowledge of births and deaths of other beings (*divyacakṣu*)
(3) Knowledge of close and remote sounds (*divyaśrotra*)
(4) Knowledge of mental states of others (*paracittajñāna*)
(5) Knowledge of the destruction of intoxicants (*āsravakṣayajñāna*) *(Liberating knowledge)*

Other powers (*ṛddhis*)

(6) Becoming light-weighted
(7) Making a distant thing proximate, a proximate thing distant.
(8) Making a small thing big, a big thing small
(9) Transformation of one thing into another
(10) Passing through a wall, mansion without sticking to it
(11) Moving in ground as if in water, in water as if on the ground
(12) Flying in the sky
(13) Touching sun/moon
(14) Bringing worlds under control

Notes

1 Here I am not presuming that Patañjali's claims for supernormal powers are "scientific". These claims are similar to the comparable claims of Buddhists and Jainas. These claims could be described as pseudo-scientific or allegedly scientific.

2 Thanks to Professor Maithrimurthi for suggesting this translation of "*saṁjñā*". The term in this context means an object of conceptualization. The translations "fusion" (*samavadhāna*) and conviction (*adhimukti*) are also due to him.

3 In this list, the eighth ideation is named as *avadātarūpanimittasaṁjñā*. But in explanation part it is termed as *avabhāsarūpanimittasaṁjñā*. So, I have accepted the latter. Similarly, in the initial list, *adhimuktisaṁjñā* and *adhimokṣasaṁjñā*, which are synonymous, are included

separately. But in the explanation part we have *vimokṣa* in place of *adhimokṣa*. *Vimokṣa* means deliverance which causes detachment (*vairāgya*).

4 Thanks to Professor Maithrimurthi who suggested to me the plausible reading as "*kauśeyaṁ vā kacaṁ vā paṭṭaṁ vā*" and suggested the possible translation of *kaca* as feather, which, too, seems plausible.

5 According to Edgerton's Buddhist Hybrid Sanskrit Dictionary, *ābhoga* means effort of focusing or earnest application of mind. The reading *vārayati* accepted by Shukla, the editor of ŚB, according to the manuscript, means that one avoids the earnest application of mind. Professor Maithrimurthi suggested to me the emendation as *dhārayati* in place of *vārayati* which is better suited to the context.

6 That is, like in the case of divine ear. The idea seems to be that one may take up a complex visible form found among people or in nature as the object of meditation but avoid any deliberate desire (*ābhoga*) towards it.

7 *Pratipadyanatā* (making efforts) is found in Edgerton's Buddhist Hybrid Sanskrit Dictionary. *Apratipadyanatā* seems to be the negation of it.

8 Here Asaṅga is probably referring to what he said in the beginning: "He applies the mind by the single pointed mental application (*samāhitabhūmikena manaskāreṇa*)".

9 Though here Asaṅga's list of the twelve *sanjñās* become complete with the last three notions, namely *vimokṣa, abhibhvāyatana* and *kṛtsnāyatana*; Asaṅga goes on adding terms like *araṇā, praṇidhijñāna* and *pratisaṁvit*. It is observed that Asaṅga has included them all in the list of special qualities (*vaiśeṣika -guṇas*) in ADS: 94–100. Asaṅga holds that these special qualities are accomplished through meditation and multiple systematic mental applications (*dhyānasanniśrayeṇa yathāvyavasthāna-manasikāra-bahulīkāratām upādāya*) (ADS: 99).

APPENDIX III

Discrepancies between Patañjali's aphorisms and Vyāsa's interpretations

In Table A3.1 I delineate some major instances where Vyāsa, in his interpretation of certain of Patañjali's aphorisms, deviates from the meaning which follows naturally from the aphorisms or which seems to be most plausible against the background of Buddhism. Some of these occurrences create a doubt in the reader's mind whether Vyāsa understood the true import of Patañjali's aphorisms. The discrepancies also suggest that the authors of *Yogasūtra* and *Yogabhāṣya* may not have been the same person, nor might they have had direct communication on the content of the text.

TABLE A3.1 Discrepancies between *Yogasūtra* and *Yogabhāṣya*

YS no.	*Patañjali's intended meaning*	*Vyāsa's discrepant interpretation*
I.8	*Viparyaya-citta-vṛtti* belongs to epistemological classification of *citta-vṛtti*s, and *avidyā* belongs to the classification of *kleśa*s. They are not identical.	Vyāsa identifies *viparyaya-citta-vṛtti* with the five-fold *avidyā*, which is the root-*kleśa*.
I.18	Patañjali describes the "other type of *samādhi*" as that with the remainder of latent impressions (*saṁskāra*). Hence, it cannot be identified with *nirbīja-samādhi*.	Vyāsa calls it *asamprajñāta-samādhi* and identifies it with *nirbīja-samādhi*.
I.23	*Īśvarapraṇidhāna* seems to mean resolve to become like *īśvara*.	Vyāsa interprets "*īśvarapraṇidhāna*" as a special kind of *bhakti* (devotion) in which the devotee surrenders all his karmas.
I.25	Patañjali describes *īśvara* as having the seed of omniscience (*sarvajñabījam*), not as someone who is omniscient.	Vyāsa interprets *īśvara* as an omniscient being which imparts knowledge out of compassion.
I.28	For Patañjali, repetition of *Om* seems to be a part of the practice called *īśvara-praṇidhāna*. It is different from *svādhyāya* (self-study).	Vyāsa identifies repetition of *Om* with *svādhyāya* (self-study).
I.33	Friendliness is about happiness (of all), compassion is about sufferings (of all), appreciative gladness is about merit (of all) and detachment is about demerit (of all).	Friendliness, compassion, gladness and indifference are to be addressed to happy, unhappy, meritorious and sinful persons respectively.
I.46	*Bīja* in *sabīja-samādhi* seems to refer to latent impression as the seed.	Vyāsa interprets *sabīja-samādhi* as the *samādhi* having external thing as the seed.
II.2	*Tanūkaraṇa* means attenuation (of *kleśa*s).	Vyāsa likens attenuated *kleśa*s with burnt seeds.
II.3	Five-fold classification of *kleśa*s seems to be based on the six-fold classification found in Buddhist Abhidharma.	Vyāsa identifies five *kleśa*s with five *viparyaya*s of Sāṅkhya.
II.4	"*Tanu*" means small .	Vyāsa associates *tanu* with the idea of "burnt seed" ("*dagdha-bīja*").
II.9	*Abhiniveśa* means dogmatic adherence to certain views.	Vyāsa interprets *abhiniveśa* as fear of death.
II.15	*Saṁskāra-duḥkha*, according to the Buddhist tradition, means unsatisfactoriness due to conditioned-ness of things. Patañjali seems to have accepted it.	Vyāsa interprets *saṁskāra-duḥkha* as the suffering caused by latent impressions.

II.27	Patañjali's concept of "*saptadhā prāntabhūmiḥ prajñā*" is a reflection of Vasubandhu's "seven knowledges which arise in the ultimate plane".	Vyāsa in his description of the seven-fold wisdom combines four Buddhist-like achievements concerning four noble truths with the three Sāṅkhya-type descriptions of liberated state.
II.32	Patañjali, by the term *svādhyāya*, seems to mean the study of emancipating sciences.	Vyāsa in fact gives two alternative interpretations of the term: (1) study of the sciences of emancipation and (2) repetition of the spell "*Oṁ*". The second interpretation does not seem to be correct.
II.46	Posture is defined as *sthirasukha* which means a steady and comfortable condition.	Vyāsa interprets *sthirasukha* not as a definition of posture in general, but the name of one of the postures. He gives a list of thirteen postures out of which *sthirasukha* is one.
II.49	Patañjali defines *prāṇāyāma* as "*gativiccheda*" of inhalation and exhalation. By the term *gativiccheda* he seems to mean distinct awareness of the motions (of inhalation and exhalation).	Vyāsa interprets *gativiccheda* as "absence of both" (inhalation and exhalation).
III.18	"By realization of impressions arises the knowledge of the previous births". Here by *saṁskāra* Patañjali seems to mean impressions (of past experiences) which cause memory.	Vyāsa interprets the term *saṁskāra* as *dharma* and *adharma* (merit and demerit), which are not directly relevant there.
III.23	By *maitryādi*, Patañjali seems to refer to all the four sublime attitudes, namely *maitrī*, *karuṇā*, *muditā* and *upekṣā*.	Vyāsa claims that *upekṣā* cannot be contemplated upon (and hence cannot generate power), which is inconsistent with his own interpretation of YS I.33.
III.54	Patañjali describes discriminating knowledge as *tāraka*, which means "saviour", that which takes to the other shore, that which liberates.	Vyāsa interprets the word *tāraka* as that which is attained through one's own intuition, and which is not based on instruction.
IV.3	By *varaṇabheda*, Patañjali seems to mean "difference in veils" which cover the created minds.	Vyāsa interprets *varaṇabheda* as "breaking the cover".
IV.12	Patañjali seems to have used the word *dharma* in "*adhvabhedād dharmāṇām*" in a typically Buddhist sense: thing or phenomenon.	Vyāsa interprets *dharma* as property by distinguishing it from property-bearer (*dharmin*).
IV.29	Patañjali says that the yogin attains *dharmamegha-samādhi* when he becomes *akusīda* even after having emancipating knowledge. The meaning of the term *akusīda is* "not-lethargic".	Vyāsa interprets the term *akusīda* as "desireless and detached".

Sources: The *Yogasūtra* aphorisms are mentioned in the first column of the table and my comments on them in the second.

GLOSSARY

This glossary contains special terms from the *Yogasūtra* as well as from Buddhist Abhidharma literature used in the book. The terms included here have been translated in different texts in varying ways. Here, the meanings given are those which are accepted in the text. Sometimes, the same term is used in different systems or contexts in different senses. Such meanings are provided by numbering them separately. Though it is a glossary of Sanskrit terms, the alphabetical order of English is followed rather than that of Sanskrit, for the readers' convenience. The diacritical marks are neglected while arranging the Romanised Sanskrit terms in alphabetical order.

abhibhvāyatana Sphere of mastery
abhijñā Supernormal knowledge
abhinirūpaṇā-vikalpa Conceptualization (or mental construction) used in fixation of the thought
abhiniveśa (1) Dogmatic view, adherence to a dogma; (2) fear of death, hope for continuation of life
abhisamaya Intuitive realization/direct realization/elucidation
ābhoga Effort, mental effort
abhyāsa Practice, repeated practice
abrahmacarya Impure conduct
adattādāna Taking what is not given, stealing
ādeśana Mind-reading
adhicittaṁ śikṣā Training of mind
adhimātra Strong
adhimokṣa Resolve, determination
adhimukti Resolve
adhipatipratyaya Governing cause

adhiprajñaṁ śikṣā Training of wisdom/insight
adhiśīlaṁ śikṣā Training of conduct
adhvan Time, temporal dimension
adhyātmaprasāda Internal tranquillity, internal serenity
adhyātma-saṃprasāda Internal tranquillity, internal serenity
āgama Scriptural authority
ahiṁsā Non-injury
āhrīkya Non-modesty
ajñāna Ignorance
ākāśānantyāyatana Sphere of unlimited space
ākiñcanyāyatana Sphere of nothingness
akliṣṭa Non-defiled
akuśala Unwholesome
akuśalamūla Unwholesome roots
Akusīda Not lethargic
akopyadharmā Unshakable one
ālambana Object
ālambanapratyaya Object as a cause (of a cognition)
ālayavijñāna Store consciousness
aliṅga Non-mark, primordial nature
anāgāmin Non-returner
anāśaya (Mind) without latent impressions
anāsrava Outflow-free, passion-free
anātma Selfless, soulless, non-self
aṅga Member, limb
aṅgamejayatva Trembling of the body
aṇimā Minuteness, capacity to become minute
anitya Impermanent
antarāya Obstacle
antardhāna Vanishing, disappearance
anuddhata(citta) Balanced, non-agitated mind
anumāna Inference
anuśāsana Instruction
anuśaya Defilement, latent defilement
anuśayin Which grows in accord with
anusmaraṇa-vikalpa Conceptualization (or mental construction) used in reflection
anutpādajñāna Knowledge of non-production
anuvyañjana Minor characteristics
anvayajñāna Subsequent knowledge
aparānta Utmost limit, death
aparigraha Non-possession
apariṇāmin Changeless
apramāda Heedfulness

apramāṇa Immeasurable
apratisaṅkhyānirodha Cessation independent of discriminative knowledge
apratisaṅkrama Without transference or transition
araṇā Absence of strife
arciṣmatī Name of the fourth bodhisattva plane
arhat Accomplished one
arthapratisaṁvit Accurate knowledge of meaning
arūparāga Lust for formless world
ārūpyasamāpatti Formless attainment
ārya-aṣṭāṅgika-mārga Noble eight-fold path
aśaikṣa One who has crossed the learner's stage
asamāhita Unstable, not-concentrated
asamayavimukta Unconditionally liberated
asaṁjñisamāpatti Attainment without ideation
asaṁskṛta Unformed, unconditioned
asaṁvara Non-restraint
āsana Posture
āsava Intoxicant
āsavakṣayajñāna Knowledge of the form, "intoxicants are destroyed"
āśaya Latent impressions of past actions
āśiṣ Hope, desire to continue one's own existence
asmitā I-notion, ego, egoism
asmimāna Ego, egoistic feeling, feeling "I am"
āsrava Influence of past actions, latent impressions of past actions
āśraya Basis, material cause
asteya Non-stealing
aśuci Impure
asukhāduḥkha-vedanā Neither pleasant nor painful sensation
atiprasaṅga Undesirable consequence
ātman Self
auddhatya Excitement, restlessness
āvaraṇa Obstruction, cover, veil
avasthā State of existence
avidyā Ignorance, misconception
avimukta Non-emancipated
avirati Non-abstinence
aviśeṣa Non-specific (general)
āyatana Sphere, sense-door
avyākṛta (1) Undefined, indeterminate; (2) unanswerable (question)
avyupaśānta Disturbed
āyuḥkarma Lifespan-determining action
bala Power
bhava Becoming, continuation of one's own existence
bhavacakra Wheel of becoming

bhāvanā Cultivation, meditative cultivation, practice
bhāvanāheya Abandonable through meditative practice
bhāvanāmārga Path of cultivation/path of meditative practice
bhāvanāmayī prajñā Insight/wisdom based on meditative practice
bhavatṛṣṇā Craving for existence
bhāvita(-citta) Contemplative mind
bhoga Experience of pleasure or pain, enjoyment
bhūmi (1) Plane, stage; (2) field, jurisdiction
bīja Seed
bodhi Awakening, enlightenment
brahmacarya (1) Continence; (2) life of purity
brahmakāyika(deva) God of Brahma's group
brahmapurohita(deva) Brahmapurohita god
caitta Mental state, mental factor
cetanā Volition
cetovimukti Emancipation of the mind
chanda/chandas Wish, liking, interest
cintāmayī prajñā Wisdom based on reflection
citi Consciousness (puruṣa)
citiśakti The power of consciousness (puruṣa)
citta Mind
cittabhūmi Mental plane
cittavṛtti Mental modification, mental state
darśana Perception, realization
darśanamārga Path of perception/realization
darśana-śakti The power of (the act of) seeing
daurmanasya Sadness, despair
dhāraṇā Fixed attention
dharma (1) Phenomenon; (2) property; (3) law
dharmajñāna Knowledge of the phenomena
dharmapratisaṁvid Accurate knowledge of factors
dharmin Property-bearer
dhātu Realm; element
dhyāna Meditation
divyacakṣus Divine eye
divyaśrotra Divine ear
dravya Substance
dṛkśakti The power by which one sees (puruṣa)
dṛṣṭadharmavedanīya (The action) the fruit of which is to be experienced in the same world/birth
dṛṣṭajanmavedanīya (The action) the fruit of which is to be experienced in the same birth
dṛṣṭi View, dogmatic view, wrong view
dṛṣṭiparāmarśa Evil adherence to wrong views

duḥkha (1) Pain, suffering, unsatisfactoriness; (2) painful, unsatisfactory
dveṣa Hatred
ekāṁśavyākaraṇīya (-praśna) (A question) which can be answered categorically
gotrakarma Action which determines birth in a particular family
grahaṇa Grasping, cognition
grahītṛ Grasper, cognisor
grāhya Object of grasping, object of cognition
guṇa Strand (of prakṛti)
guṇādhikāra Operation of the strands (of prakṛti)
hāna Abandonment
hānopāya The means to the abandonment
hetu Cause
hetu-pratyaya Main cause, efficient cause
heya To be abandoned
heyahetu The cause of what is to be abandoned
indriya (1) Faculties as six sense faculties; (2) faculties as five means to spiritual attainment
īryāpatha Manner of behaviour, deportment
īśitṛtva Capacity to create or destroy
īśvara (1) Supreme being; (2) God as creator
īśvarapraṇidhāna Resolve regarding the supreme being
janmakathantāsambodha Knowledge about "how" of one's births
jāti Birth, life-state
jīvitendriya Vital faculty, life principle
jñāna Knowledge
jñeyāvaraṇa The veil (of ignorance) constituted by the knowable
jyotiṣmatī (Mental state) which is full of luminosity
kaivalya (1) Isolation, emancipation; (2) seclusion, separation
kāmacchanda Excitement for sensuous pleasure
kāmadhātu The world (realm) of desire for sensuous pleasure
kāmarāga Lust for sensuous pleasure
kāmāvacara-citta Mind dwelling in the realm of desire
kāritra Activity
karuṇā Compassion
kaṣāya Impurity
kausīdya/kauśīdya Indolence, lethargy, inertia
khyāti Discriminative knowledge
kleśa Defilement, affliction
kleśāvaraṇa The veil constituted by defilements
kliṣṭa Defiled
kriyāyoga Spiritual discipline of action
krodha Anger
kṛtaparicaya Experienced practitioner
kṛtaparijaya One who has mastered the method

kṣānti Patient acceptance, forbearance
kṣayajñāna Knowledge of extinction
kṣipta Unstable, disturbed
kūrmanāḍī Bronchial tube
kuśala Wholesome, moral
kuśalamūla Wholesome root
kusīda Indolent, lethargic
laghimā Lightness, capacity to become light
lakṣaṇa (1) Characteristic; (2) temporal characteristics
laukikamārga The worldly path
leśyā Colour of a soul caused by the colour of karma attached to it
līna(-citta) Indolent (mind)
lobha Covetousness
lokottara Supra-mundane
madhyastha Situated in the middle, equanimous
mahābhūmika Of great extent (mental factors belonging to all stages)
mahadgata(-citta) Lofty (mind)
mahāvidehā (vṛttiḥ) Great disembodied state
mahāvrata Great vow
mahimā Largeness, the capacity to become large
maitrī Friendliness, loving kindness
mala Stain, dirt, impurity
māna Conceit
manas Mind
manasikāra/manasikāra Attention, mental application, thought
manojavitva Mind-like speediness
mantra Spell
mārga Path
mati Thought
middha Sleepiness, drowsiness
moha Foolishness mṛdu Weak, mild
muditā Joy, appreciative joy
mūla Root
mūrdhajyotiṣ Light in the head
naiva-saṃjñā-nāsaṃjñāyatana The sphere of neither perception nor non-perception
naraka Hell
nirbījasamādhi Meditative absorption without seed
nirmāṇacitta Magically created mind
nirodha Cessation
nirodha-samāpatti Attainment of cessation
nirbīja-samādhi Accurate knowledge of explanations
niṣevaṇabhāvanā Contemplation of maintaining
nivaraṇa/nīvaraṇa Hindrance

niyama Observance
paracittajñāna Knowledge of the thoughts of others
paramāṇu Atom
paranirmitavaśavartin (Gods) controlling (enjoyments) magically created by others
pariṇāmaduḥkha That which results into suffering
paripṛcchya-vyākaraṇīya(-praśna) (A question) which can be answered after counter-questioning
parītta (citta) Narrow (mind), mean (mind)
phala Result, fruit
prabodha Awakening
pracchardana Expulsion (of breath)
pragṛhīta (citta) Energetic (mind)
prajñā Wisdom
prajñapti-sat Real by designation only (not substantially real)
prajñāvimukti Emancipation of wisdom
prākāmya Non-frustration of desire
prakṛti (1) Nature; (2) primordial nature
prakṛtilaya Beings merged in their original nature
pramāda Non-diligence, heedlessness
prāmodya Joy
prāṇāyāma Breath-regulation
praṇidhāna Resolve
praṇidhijñāna Knowledge resulting from resolve
prāntabhūmi-prajñā Ultimate wisdom, wisdom belonging to the highest plane
prāpti (1) Attainment, (2) capacity to reach (touch) a remote object
praśrabdhi Peace, repose, calm
prasupta (kleśa) Dormant (defilement)
prātibha(-jñāna) Intuitive knowledge
pratibhānapratisaṁvit Accurate intuitive knowledge
pratigha Hatred, hostility
prātihārya Marvel
pratilambha-bhāvanā Contemplation of obtaining
pratipakṣa Opposite approach, antidote
pratipakṣabhāvana/pratipakṣabhāvanā Contemplation of the antidote
pratiprasava Reverse creation, reverse generation
pratisaṁvid Accurate knowledge
pratyaya (1) Cognition, experience; (2) causal condition
pratyāhāra Withdrawal of senses
preta Hungry ghost
pudgala (1) Person, self; (2) matter
puṇya Merit
puruṣa Spirit
rāga Desire, attachment

raṇa Strife
ṛddhi Supernormal power
ṛtambharā prajñā Truth-bearing wisdom
rūpa (1) Matter; (2) form
rūpadhātu The world (realm) of form
rūparāga Lust for the world of forms
rūpāvacara (Gods) belonging to the realm (world) of form
sabīja-samādhi Meditative absorption with seed
samādhi Concentration, meditative absorption
samādhibhāvanā Developing concentration
samāhita(-citta) Concentrated, stable (mind)
samanantara Immediate (immediately preceding or succeeding)
samanantarapratyaya Immediately preceding causal condition
samāpatti (1) Attainment (of a state of meditation); (2) attainment (of an object through meditation)
śamatha (1) Peace; (2) a kind of meditation, concentration meditation
saṁjñā Perception
saṁjñāveditanirodha Cessation of perception and feeling
samprajanya Mindfulness, circumspection
samprajñāna Mindfulness, circumspection
sampramoṣa Wiping, cleaning, removal
saṁskāra (1) Conditioned thing; (2) formation; (3) latent impression of an action; (4) impression of a cognition
saṁskāra-duḥkha Painful due to conditioned-ness
samudaya Origination, cause
saṁvara Restraint
saṁvṛtijñāna The knowledge of all objects
saṁyama Integrated concentration
saṁyojana Fetter
saṅghāta Collection, combination
saṅkara Admixture
saṅkṣipta (citta) Internally but temporarily focused (mind)
santoṣa Contentment
santuṣṭi Contentment
sarvārthatā Dispersiveness, attention to everything
sāsrava(-citta) (Mind) with latent impressions, impure mind
śāśvatadṛṣṭi View of eternity
satkāyadṛṣṭi Dogmatic belief in self/soul
sattvapuruṣānyatākhyāti Knowledge of the distinction between mind and spirit
satya (1) Truth; (2) truthfulness
śauca/śauceya Cleanliness, purification
saumanasya Gladness
siddha Perfect being
siddhi Supernormal achievement

śīla Morality
śīlavrataparāmarśa Attachment to mere rules and rituals
smṛti (1) Mindfulness; (2) recollection
smṛti-pariśuddhi Purification of mindfulness
smṛtyupasthāna Application of mindfulness
sparśa Contact
śraddhā Faith, confidence based on knowledge
śrāvaka Disciple
śruta Learning
sthāpanīya(-praśna) (A question) to be set aside
Sthūla Coarse, gross
Styāna Sloth, torpor
śubha Pure
śuci Pure, clean
sukha (1) Happiness; (2) pleasant
sūkṣma Subtle
sva Owned thing, belonging
svabhāva-vikalpa Natural mental construction
svādhyāya Self-study
tanu Small
tanūkaraṇa Attenuation
tāpaduḥkha Suffering, which is pain by nature
tapas Austerity, penance
tāraka Liberating one
trāyastriṁśa(deva) The thirty-three (gods)
tṛṣṇā Craving
uccheda Annihilation
udāra Gross
uddhata Excited, agitated, restless
upādāna Clinging
upakleśa Secondary affliction
upapatti Birth
upāyāsa Irritation
upekṣā (1) Even-mindedness, equanimity; (2) indifference
upekṣāpariśuddhi Purification of equanimity
vairāgya Detachment, renunciation
vaiśāradya (1) Fearlessness, confidence; (2) proficiency, serenity
vajrasaṁhananatva Thunderbolt-like hardness
vāsanā Latent mpression, flavour, traces
vaśitva Control over things
vastu Entity, object
vedanā Feeling
vedanīya-karma The action that determines the experience of pleasure and pain
vibhajyavyākaraṇīya (-praśna) (A question) which can be answered after splitting it (into alternatives)

vicāra (1) Subtle thought, (2) a subtle object of thought
viccheda Differentiation
vicchinna Divided, separated, broken out, differentiated, manifested
vicikitsā Doubt
videha Disembodied being, god
vidhāraṇa Retention (of breath)
vihiṁsā Harmfulness
vijñāna Consciousness
vijñānānantyāyatana Sphere of infinite consciousness
vikalpa Mental construction
vikaraṇabhāva Capacity to control things
vikṣepa Distraction
vikṣipta (-citta) (1) Distracted (mind); (2) externally focused (mind)
vimokṣa Deliverance
vimukta(-citta) Emancipated (mind)
vimukti Emancipation
vipāka Maturation, fruition
viparyāsa Perversion
viparyaya Error, false cognition, illusory cognition
vipaśyanā Insight, insight meditation
vīrya Energy, vigour
viśeṣa (1) Particular (2) specific characteristic
viśokā Sorrowless
vītarāga Attachment-free person
vitarka (1) Gross thought; (2) gross object of thought; (3) unwholesome thought
vivekajaṁ jñānam (1) Discriminative knowledge; (2) knowledge arising from dissociation
vṛtti State, manifestation, function, modification
vyāpāda Hatred
vyāyāma Exertion
vyupaśānta Undisturbed, peaceful
vyutthāna Outward-directedness, extravert-ness, coming out of meditative trance
yama Restraint
yatrakāmāvasāyitva Fulfilment of desires
yoga (1) Attachment; (2) self-disciplining aimed at a spiritual goal; (3) cessation of mental states

AIDS TO READING ROMANIZED SANSKRIT

Roman letters (with diacritical marks) standing for *Devanāgarī* letters

Romanization	Devanāgarī	*Romanization*	Devanāgarī	*Romanization*	Devanāgarī
A, a	अ	Ḥ, ḥ	ः	P, p	प
Ā, ā	आ	I, i	इ	Ph, ph	फ
Ai, ai	ऐ	Ī, ī	ई	R, r	र
Au, au	औ	J, j	ज	Ṛ, ṛ	ऋ
B, b	ब	Jh, jh	झ	Ṝ, ṝ	ॠ
Bh, bh	भ	K, k	क	S, s	स
C, c	च	Kh, kh	ख	Ś, ś	श
Ch, ch	छ	L, l	ल	Ṣ, ṣ	ष
D, d	द	ḷ	ऌ	T, t	त
Dh, dh	ध	M, m	म	Th, th	थ
ḍ	ड	Ṁ, ṁ	ं	ṭ	ट
ḍh	ढ	N, n	न	ṭh	ठ
E, e	ए	Ṅ, ṅ	ङ	U, u	उ
G, g	ग	Ṇ, ṇ	ण	Ū Ū	ऊ
Gh, gh	घ	Ñ, ñ	ञ	V, v	व
H, h	ह	O, o	ओ	Y, y	य

BIBLIOGRAPHY

Primary sources

ADS: *Abhidharmasmuccaya*, Edited by Prahlad Pradhan, Visva Bharati, Santiniketan, 1950.

AK: *Abhidharmakośa* as in *Abhidharmakośabhāṣya of Vasubandhu*, Edited by Professor P. Pradhan, K. P. Jayaswal Research Institute, Patna, 1967.

AKB: *Abhidharmakośabhāṣya* as in Professor P. Pradhan, 1967.

BG: *Śrīmad-bhagavadgītā* (The Vedānta Text), Edited by J.L. Bansal, JPH, Jaipur, 2013.

BSB: *Bodhisattvabhūmi [Being the XVth Section of Asaṅgapāda's Yogācārabhūmi]*, Edited by Nalinaksha Dutt, Vol. 7, K. P. Jayaswal Research Institute, Patna, 1966.

CS: *Carakasaṁhitā by Agniveśa with the Commentary by Cakrapāṇidatta*, Edited by J.T. Acharya Vaidya, Nirnay Sagar Press, Bombay, 1941 (3rd Edition).

DBI: *Daśabhūmīśvaro nāma mahāyānasūtram*, Edited by Ryūkō Kondō, the Darjyō Bukkyō Kenyō-kai, Tokyo, 1936.

Dhammapada, Edited by S. Sumangala Thera, PTS, London, 2014.

GS: *Gheraṇḍasaṁhitā*, The Original Sanskrit and an English Translation by James Mainson, YogaVidya.com, Woodstock, NY, 2004.

HP: *Haṭhpradīpikā of Svātmārāma*, Edited by Swami Digambarji, Kaivalyadham Samiti, Lonavla, 2018.

Kaṭhopaniṣad as Included in *Upaniṣads*.

Māṇḍūkyopaniṣad as Included in *Upaniṣads*.

Mokṣadharma as Included in *The Śāntiparvan (Part III: Mokṣadharma, A)*, Edited by S.K. Belvalkar, Bhandarkar Oriental Research Institute, 1954.

NBV: *Nāgojibhaṭṭavṛtti* as in *The Yogadarśana of Patañjali with the Commentaries Bhāvāgaṇeśīya and Nāgojībhaṭṭīya*, Edited by Mahadeva Gangadhar Bakre, Nirnay Sagar Press, Bombay, 1917.

PB: Praśastapāda's Commentary of *Vaiśeṣikasūtra* as Included in *Vaiśeṣikadaśanam*, Edited by Shrikrishna Shastri, Nirnay Sagar Press, Mumbai, 1946.

Praśnopaniṣad as Included in *Upaniṣads*.

RMV: *Rājamārtaṇḍavṛtti* as in *Pātañjalayogasūtraṁ Bhojadevakṛta-rājamārtaṇḍavṛttisametam*, Edited by Ram Shankar Bhattacharya, Bharatiya Vidya Prakashan, Varanasi, 1997.

ŚB: *Śrāvakabhūmi of Ācārya Asaṅga*, Deciphered and Edited by Karunesh Shukla, K. P. Jayaswal Research Institute, Patna, 1973.

SK: *Sāṅkhyakārikā of Īśvarakṛṣṇa (with Sāṅkhyatattvakaumudī of Vācaspatimiśra)*, Jaya Krishna das Haridas G, Benares, 1937.

TAS: *Tattvārthasūtra* as in Tatia, 2007.

TV: *Tattvavaiśāradī* of Vācaspatimiśra as in YD.

Upaniṣads: *The Upanishads*, Edited by V.P. Vaidya, Nirnay Sagar Press, Bombay, 1932.

UV: *Udānavarga*, Edited by Franz Barnhard, Götingen, 1965.

VB: *Vyāsabhāṣya on Yogasūtra* as Included in YD.

Vi: *Viṁśatikā* as in *Vijñaptimātratāsiddhi (Prakaraṇadvayam) of Ācārya-Vasubandhu*, Edited by Ram Shankar Tripathi, Sampurnanand Sanskrit University, Varanasi, 1992.

VP: *Vākyapadīyam*, Edited by Baladeo Upadhyay, Varanaseya Sanskrit Vishvavidyalay, Varanasi, 1968.

YAB: *Yogācārabhūmi of Ācārya Asaṅga*, Edited by Vidhushekhar Bhatacharya, University of Calcutta, 1957.

YB: *Yogabhāṣya (Vyāsabhāṣya)* as Included in YD.

YBV: *Yogasūtra-bhāṣyavivaraṇa of Śaṅkara*, Vol. 1 and Vol. 2, Edited and Translated by T.S. Rukmani, Munshiram Manoharlal Publishers Pvt. Ltd., New Delhi, 2001.

YD: *Pātañjalayogadarśanam with Tattvavaiśāradī of Vācaspatimiśra, Yogavārtika of Vijñānabhikṣu and Vyāsabhāṣya*, Edited by Shri Narayan Mishra, Bharatiya Vidya Prakashan, Varanasi, 1971.

YS: *Yogasūtra* of Patañjali as Included in YD.

YS(V): *Yogasūtra* Version Accepted in YBV.

YV: *Yogavārtika of Vijñānabhikṣu* as Included in YD.

Secondary sources

Abhyankar, V., "Upodghāta", as Included in *Yogasūtras of Patañjali*, Bhandarkar Oriental Research Institute, Pune, 2006, pp. XIII–XXIII.

Apte, V.S., *The Practical Sanskrit English Dictionary*, Prasad Prakashan, Poona, 1957.

Bahulkar, S.S., "On the Nine Categories of Yogin (Mentioned by Commentaries of Yogasūtras I.21–22)", *Proceedings of the All India Oriental Conference*, 32nd Session, Ahmedabad, 1985, pp. 443–450.

Bhave, S.M., "Bauddhamata va patañjalichā yoga (Marathi Introduction to La Vallee Poussin, 1937)", *Parāmarśa*, 13.4 (February 1992), pp. 1–17.

Brahmalīnamuni, S., Ś. (Ed.), *Hindī Pātañjalayogadarśana*, Chaukhambha Prakashan, Varanasi, 2014 (Reprint).

Bronkhorst, J., "Patañjali and the Yoga Sūtras", published in *Studien zur Indologie und Iranistik*, 10 (1984 [1985]), pp. 191–212.

Bronkhorst, J., *The Two Traditions of Meditation in Ancient India*, Motilal Banarsidass, New Delhi, 1993 (2nd Edition).

Bronkhorst, J., "Yoga and Seśvara Sāṁkhya", published in *Journal of Indian Philosophy*, 9 (1981), pp. 309–320.

Chatterjee, A.K., *Readings in Yogācāra Buddhism*, Centre of Advanced Studies in Philosophy, Banaras Hindu University, Varanasi, 1971.

Dasgupta, S., *A History of Indian Philosophy*, Vol. 1, Motilal Banarsidass, New Delhi, 1975 (1st Indian Edition).

Davids Rhys, T.W. and Stede, W., *Pali English Dictionary*, Motilal Banarsidass, New Delhi, 1997 (Reprint).

Dhammajoti, B., *Sarvāstivāda Abhdharma*, Centre for Buddhist Studies, the University of Hong Kong, 2009.

Edgerton, F., *Buddhist Hybrid Sanskrit Grammar and Dictionary, Volume II: Dictionary*, (First edition, New Haven, 1953), Motilal Banarsidass, Delhi, 1993 (Fifth Indian Reprint).

Feuerstein, G., *The Yogasūtra of Patañjali: An Exercise in the Methodology of Textual Analysis*, Gulab Vazirani for Arnold-Heinemann, New Delhi, 1979.

Frauwallner, E., *History of Indian Philosophy* (Translated from Original German Into English by V.M. Bedekar), Vol. 1, Motilal Banarsidass, New Delhi, 1984 (Reprint).

Glasenapp, H.V., *Doctrine of Karman in Jaina Philosophy*, P. V. Research Institute, Varanasi, 1942.

Gokhale, P.P., "Ethics in Jaina Philosophical Literature: The Doctrine of Ahimsa", in Rajendra Prasad (Ed.) published in the Vol. *A Historical-Developmental Study of Classical Indian Philosophy of Morals*, Under PHISPC Project, Concept Publishing Company, New Delhi, 2009.

Gokhale, P.P., *Inference and Fallacies Discussed in Ancient Indian Logic (With Special Reference to Nyāya and Buddhism)*, Sri Satguru Publications, New Delhi, 1992.

Gokhale, P.P., *Lokāyata/Cārvāka: A Philosophical Inquiry*, Oxford University Press, New Delhi, 2015.

Gokhale, P.P., "Omniscience", in K.T.S. Sarao and Jeffery D. Long (Eds.) *Encyclopedia of Indian Religions*, Vol. 2 (Series Editor: Arvind Sharma), Springer, 2017, pp. 813–820.

Gokhale, P.P., "Re-Understanding Indian Moral Thought", in S.E. Bhelke and P.P. Gokhale (Eds.) *Studies in Indian Moral Philosophy, Problems, Concepts and Perspectives*, IPQ Publication Pune University, Pune, 2002.

Gokhale, P.P., "Working of Mind in *Sati* Meditation: Some Issues and Perspectives", in Kalpakaṃ Sankarnarayan and others (Eds.) *Dharma and Abhidharma*, Vol. 2, Somaiya Publications, Mumbai, 2007.

Harimoto, K., "Review: *Yogasūtra-bhāṣyavivaraṇa of Śaṅkara* by T. S. Rukmani published by Munshiram Manoharlal Publishers", *Journal of the American Oriental Society*, 124.1 (January 2004), pp. 176–180.

Hirakawa, A., *Index to the Abhidharmakośabhāṣya (P. Pradhan Edition)*, Part One, Daizo Shuppan Kabushikikaisha, Tokio, 1973.

Iyengar, B.K.S., *Light on Yoga*, George Allen and Unwin Ltd., London, 1965.

Jayatilleke, K.N., *Early Buddhist Theory of Knowledge*, Motilal Banarsidass, New Delhi, 2010 (Reprint).

Larson, G.J., "An Old Problem Revisited: The Relation between Sāṅkhya, Yoga and Buddhism", *Studian zur Indologie und Iranistik*, 15 (1989), pp. 129–146.

Larson, G.J. and Bhattacharya, R.S., *Yoga: India's Philosophy of Meditation: Encyclopedia of Indian Philosophies*, Vol. 12, Motilal Banarsidass, New Delhi, 2008.

La Vallée-Poussin, L. de "Le Bouddhisme et le Yoga de Patañjali", published in *Melanges Chinois et Bouddhiques*, 5 (1936–37 volume), 1937, pp. 223–242.

Mass, P., "A Concise Historiography of Classical Yoga Philosophy", in Eli Franco (Ed.) *Periodization and Historiography of Indian Philosophy*, the De Nobili Research Library, Wein, 2013.

Monier-Williams, S.M., *Sanskrit-English Dictionary*, Munshiram Manoharlal, New Delhi, 1976.

Narada, *The Buddha and His Teachings*, Buddhist Missionary Society, Malaysia, 1988 (4th Edition).

Rhys Davids, T.W. and Stede, W., *Pali English Dictionary*, Motilal Banarsidass Publishers Pvt. Ltd., New Delhi, 1997 (Reprint).

Rukmani, T.S. (Ed. and Trans.), *Yogasūtra-bhāṣyavivaraṇa of Śaṅkara*, Vol. 1 and Vol. 2, Munshiram Manoharlal Publishers Pvt. Ltd., New Delhi, 2001.

Schmithausen, L., *Ālayavijñāna: On the Origin and the Early Development of a Central Concept of Yogācāra Philosophy*, The International Institute of Buddhist Studies, Tokyo, 1987.

Stcherbatsky, *The Central Conception of Buddhism*, Sushil Gupta (India) Ltd., Calcutta, 1961 (3rd Edition).

Taimni, L.K., *The Science of Yoga*, The Theosophical Publishing House, Madras, 1965.

Tandon, S.N., *A Re-Appraisal of Patañjali's Yogasūtras in the Light of the Buddha's Teaching*, Dhamma Books, Igatpuri, 1995 (1998 Reprint).

Tatia, N., *Umāsāti/Umāsvāmi's Tattvārthasūtra: That Which Is*, Motilal Banarsidass, New Delhi, 2007.

Thomas, E.J., *The Life of Buddha: As Legend and History*, Motilal Banarsidass, New Delhi, 2005 (Reprint).

Vivekananda, S., *Rajayoga or Conquering the Internal Nature*, Advaita Ashram, Mayavati, Pithorgarh, Himalayas, 1982.

Woods, J.H. (Translated and Annotated), *The Yogasūtra of Patañjali*, Dover Publications, Mineola, NY, 2003.

Wujastyk, D., "Some Problematic Yogasūtras and Their Buddhist Background", in Philipp Maas et al. (Eds.) *Yoga in Transformation*, Vienna University Press, 2016.

Yamashita, K., *Pātañjala Yoga Philosophy with Reference to Buddhism*, Firma KLM Pvt. Ltd., Calcutta, 1994.

Yardi, M.R., *Yoga of Patañjali*, Bhandarkar Institute, Poona, 1979.

INDEX

Abhidharma 12, 24, 42, 62, 127, 178
abhijñā 128, 145, 149, 152–153, 212, 217
abhyāsa 31–33, 196
adhyātmaprasāda 59, 202
adhyātmasamprasāda 59, 201
akusīda 180, 198, 221
ālayavijñāna 58, 62, 176
ānāpāna 105, 107, 108–109, 140
anuśaya 73, 196; *bhāvanāheya* 75, 76; *dṛgheya* 75, 76; *see also kleśa*
apramāṇa see immeasurables
arūpadhyāna 35–36, 61; *see also ārūpya*
ārūpya 203–204
ārūpya-samāpatti 200
asmitā 34–35, 70–71, 72–73, 164–165, 193, 195
avidyā 29, 70–72, 85, 189, 190, 193

bhūmi 21, 98, 120, 179, 208–210
bodhisattva: *bodhisattava*'s journey 177, 179–181, 182, 207–210; *bodhisattva*'s *praṇidhāna* 39
Bronkhorst, Johannes 8, 9, 10, 17, 63, 66
Buddhist idealism 171, 175–176; *see also* Idealist Buddhist

Carakasaṁhitā 6, 48, 132
cause: types of 88–91, 168–169
cetovimukti 87–88
citta 21–24; *akliṣṭa* 26; *anāsrava* 26; *ekāgra* 22; *kliṣṭa* 21, 26; *kṣipta* 21–22; *kuśala* 21; *niruddha* 22–23; *sāsrava* 26; *vikṣipta* 21–22
cittabhūmi 21–23
citta-vṛtti 24–25, 28–31
cosmology: Brahmanical 135; Buddhist 135
cosmos: Vasubandhu's picture of 135–139; Vyāsa's picture of 135–139

Dasgupta, Surendranath 7, 8
defilements *see kleśa*
detachment *see vairāgya*
dhāraṇā 117, 119, 120, 121, 203
Dharmakīrti 28, 60, 129
dharmameghā bhūmi 179–180, 207–210
dharmamegha-samādhi 179–180, 181–182, 195, 208
dhyāna 118, 119, 20–23; *samāpatti* 200; *upapatti* 200; *see also arūpadhyāna*; *rūpadhyāna*
Diṅnāga 28, 60, 129, 176
divine ear 214–215; *see also divyaṁ śrotam*
divine eye 215; *see also divyacakśu*
divyacakśu 134; *see also* divine eye
divyaṁ śrotam 145; *see also* divine ear
dveṣa 70–71, 74, 98

fetters *see saṁyojana*
Feuerstein, Georg 10
Frauwallner, Erich 10, 63, 66, 160, 185, 198

Haṭhayoga 1, 2, 4, 50, 102, 103
Haṭhayogic concept of *prāṇāyāma* 50, 103, 104

idealist Buddhist 172, 173; *see also* Buddhist idealism
immeasurables 47, 132, 197

īśvara 3, 39–41, 191, 193
īśvarapraṇidhāna 39–40, 42, 69, 101, 196

Jaina: classification of actions 166; concept of knowledge covering karma 108; theory of karma 77–78
Jainas on extra-sensory perception 134
Jainism 27, 92–94, 147, 181–182

kaivalya 25, 75, 85, 184, 191, 195
kāritra 124, 125, 126, 170, 194
karma 131, 147, 180–181; classification of 76–77, 166
karmasthāna 51, 52
karuṇā 46–48, 132–133, 197
kausīdya/kauśīdya 21, 180, 186
kleśa 40, 70, 75–77; classification of 70–71; *dhyānaheya* 76; dormant 70; *pratiprasavaheya* 75–76; *see also anuśaya*
knowledge: of other minds 130; of other's mental states 215; *see also paracittajñāna*

Larson, Gerald 9, 10, 17
La Vallee Poussin 10, 11, 110, 111, 112, 158, 160

Mahāyāna 98, 177, 179, 181, 195, 208
maitrī 46–49, 132–133, 197
Mass, Philipp 9, 17
mind: Vasubandhu's classification of 21
momentariness 174, 183
muditā 46–49, 132–133, 197

nirmāṇacitta 163, 164, 165
nirodha-samāpatti 22–23, 122, 197
nirvāna 25, 85

omniscience 40–41, 152–153, 181, 197

paracittajñāna 87, 130; *see also* knowledge, of other's mental states
pariṇāma 123–125, 127, 163, 194
pariṇāmavāda 124–125, 194
Patañjali 5–6; as the author of *Vyākaraṇamahābhāṣya* 7; as the author of *Yogasūtra* 5, 7; period of 5–8
Patañjalis: three 6; two 7–8
prajñā 59–61; *bhāvanāmayī* 60, 177; *ṛtumbharā* 59–61, 92, 120, 177, 202; as sevenfold wisdom 86–88, 196
prajñāvimukti 87–88
prāṇāyāma 103–108; as *ānāpānasmṛti* 105, 109; in Haṭhayogic sense 103–104
praṇidhāna 39, 196
pratībhānapratisaṁvid 141
pratipakṣabhāvana/pratipakṣabhāvanā 95–98
pratiprasava 75, 191
pratyāhāra 89, 109

rāga 73; *arūpa* 61; *rūpa* 61
Rājayoga 1, 2, 4
ṛddhi 149, 162, 163, 212–218
rūpadhyāna 34, 36, 61, 117, 203; stages of 34, 201

samādhi 89, 91, 118–119, 200–203; *asamprajñāta* 36, 196, 204; *nirbīja* 36, 62, 92, 121, 196–197; *sabīja* 58, 92, 196, 197; *samprajñāta* 34–36, 196, 197
samāpatti 54, 55, 196–202; according to Vasubandhu 54; *nirvicāra* 55–57, 203; *nirvitarka* 55, 57; *savicāra* 55, 57, 203; *savitarka* 55, 57
saṁskāra 31, 36, 61–62, 129–130, 167–168
saṁyama 119, 197, 212
saṁyojana 42, 61, 204
Śaṅkara 3, 16, 73–74
Sarvāstivāda 8, 10, 124, 170, 194
Sautrāntika 10, 125, 128, 129, 173, 178
śīla 91, 92
smṛti 30–31, 43; *-pariśuddhi* 56, 201–202
sublime attitudes 46; *see also* immeasurables
svādhyāya 42, 69, 94–95, 101

Tandon, S.N. 65, 67, 104, 107, 132, 143, 185
temporal dimensions 124, 125, 126, 169–170

Umāsvāti 43, 92–93
Upaniṣad 11, 28, 41, 205; *Kaṭha* (*Kaṭhopaniṣad*) 109; *Māṇḍūkya* (*Māṇḍūkyopaniṣad*) 28, 65
upekṣā 46–49, 133, 197–198

Vaibhāṣika 129, 170, 178, 179
vairāgya 31–32, 33, 153; *adhobhūmi* 32, 61; *laukika* and *lokottara* 153, 196; lower and higher 33–34, 153, 196; *see also* detachment
Vaiśeṣika 27, 145–146, 148
vāsanā 58, 62, 167, 176, 181, 185, 189
vicāra 34–35, 117–118, 200–202
vikalpa 7, 27, 29–30, 129; Asaṅga on 30
viparyāsa 72, 195
viparyaya 29; according to Sāṅkhya 27; according to Vaiśeṣika 27
vipaśyanā/vipassanā 67, 199, 210
vitakka 12, 117, 203
vītarāga 51, 87
vitarka 30, 34–35, 95–97, 117, 200–203; *akuśala* 96, 97
vyutthāna 26, 121–122, 143

Woods, James Haughton 7, 17, 63

Yoga 1, 2, 20, 23, 206; Asaṅga's conception of 205–207; eight-limbed 89; Pātañjala 1, 2, 4, 196–198
Yogabhāṣya 8–10; the author of 8–10
Yogācāra 10–11, 12, 58, 62, 176, 205, 207; Early 58, 62, 188
Yogasūtra 5; the author of 5–6; the period of 6–8; and *Yogabhāṣya* 8–10

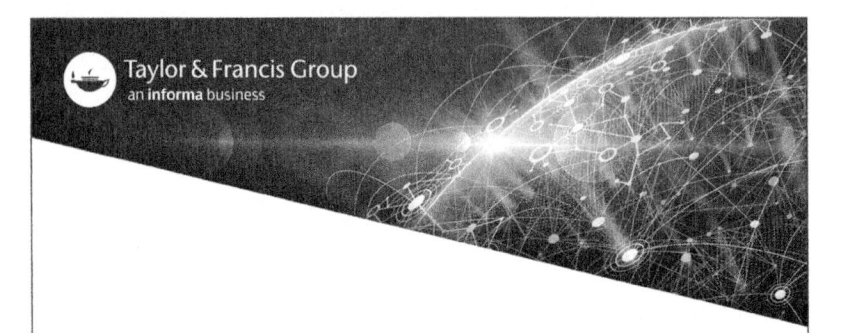

Printed in Great Britain
by Amazon

79246409R00142